For Grade **5**

Student Workbook

This Book Belongs To:

Science on Target

Using Graphic Organizers to Improve Science Skills

Written By:
Andrea Karch Balas, Ph.D

Show What You Know ®
Publishing

Published By:

Show What You Know® Publishing

A Division of Englefield & Associates, Inc.

P.O. Box 341348

Columbus, OH 43234-1348

1-877-PASSING (727-7464)

www.showwhatyouknowpublishing.com

Printed in the United States of America

11   20 19 18 17 16 15 14 13 12 11 10 9 8 7 6 5 4 3

ISBN: 1-59230-332-3

About the Author

Andrea Karch Balas, Ph.D., is an educator and a scientist who has taught both in the traditional classroom and in nonformal educational settings, from kindergarten to adult. Andrea has presented her research on the teaching and learning of Science both nationally and internationally. Andrea recevied her doctorate in education from The Ohio State University and is currently the General Curriculum Coordinator in a private K–12 school. In addition to this book, Andrea is the co-author of the Ready, Set, Show What You Know® series for grades K–3 in Ohio and Florida.

Acknowledgements

Show What You Know® Publishing acknowledges the following for their efforts in making these skill-building materials available for students, parents, and teachers.

Cindi Englefield, President/Publisher
Eloise Boehm-Sasala, Vice President/Managing Editor
Jill Borish, Production Editor
Christine Filippetti, Production Editor
Jennifer Harney, Illustrator/Cover Designer

 © Englefield & Associates, Inc.

Table of Contents

Visual Glossary of Graphic Organizers

A graphic organizer is an instructional tool used to illustrate prior knowledge about a topic. This visual glossary will give you an idea of what the graphic organizers used in the *Science on Target for Grade 5, Student Workbook* look like, as well as the best way for you to put them to practical use in your everyday Science lessons.

Compare and Contrast Chart

	name 1	name 2
attribute 1		
attribute 2		
attribute 3		

Compare and Contrast Chart uses: Show similarities and differences between two things (people, places, events, ideas, etc.).

Examples include: The comparison of plants and animals. Plants and animals are eukaryotes but plant cells contain additional organelles (cell walls, chloroplasts). Plants make their own food through the process of photosynthesis. Plants are the beginning of the food chain for all animals.

Questions to ask: What are the objects, processes, or procedures being compared? What are the component parts of each? How are they similar? How are they different?

Concept Map

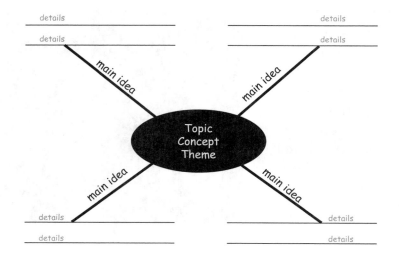

Concept Map uses: Description of a central idea and the relationship of supporting ideas, topics, or functions.

Examples include: If "States of Matter" is the central topic, the connected ideas would be solid state, liquid state, and gaseous state. For each of the states of matter, details could be added to relate the form of the molecules in the state. For example, solids are rigid around a fixed point, and liquids and gases take the shape of their containers.

Questions to ask: How are the ideas connected or interrelated? What details are important to include on the map?

Cycle

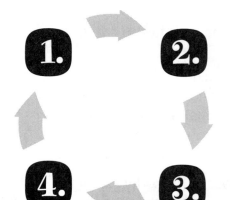

Cycle uses: Illustrates the repeated stages or events that occur to create a specific product or event.

Examples include: The illustration of cell reproduction, rock formation, and weather conditions.

Questions to ask: What are the critical stages or events in the cycle? What is the sequence in the activity? Where or what is the event that promotes the continuity of the cycle?

© Englefield & Associates, Inc.

Diagram

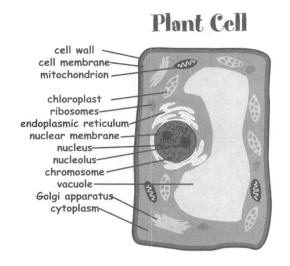

Plant Cell

cell wall
cell membrane
mitochondrion

chloroplast
ribosomes
endoplasmic reticulum
nuclear membrane
nucleus
nucleolus
chromosome
vacuole
Golgi apparatus
cytoplasm

Diagram uses: Provides a tool for describing the relationship between the parts of a system.

Examples include: The model of the cell, the layers of soil, or the components of a circuit.

Questions to ask: What are the parts of the object, theme, or system? What is a visual representation of the information about this process or experiment?

Dichotomous Key

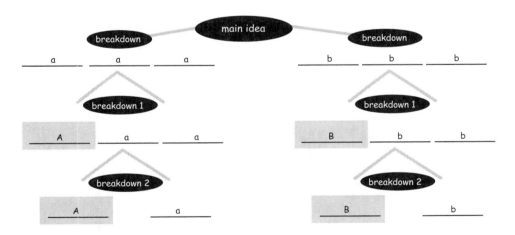

Dichotomous Key uses: Characteristics are used to divide a group of objects, organisms, or ideas into two groups until one object remains in each division.

Examples include: Separating organisms into Kingdoms based on attributes.

Questions to ask: How can I separate these objects, organisms, or ideas based on a given trait?

Graph

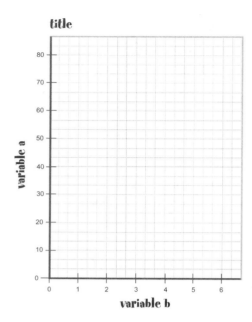

Graph uses: To visually depict collected data.

Examples include: The results of a survey or changes that occur in a given situation.

Questions to ask: Vary depending on the graph, e.g., How many people like a particular fruit? How did the temperature change over time?

Organizational Outline

Title

I. Topic 1
    A. detail
    B. detail

II. Topic 2
    A. detail
    B. detail

Organizational Outline uses: Organizing information, sequencing processes or events.

Examples include: Organizing information from a Science article or presenting information about a mutlifaceted topic such as ecosystems.

Questions to ask: What details are given to support the topic?

       © Englefield & Associates, Inc.

Series of Events Chain or Flow Chart

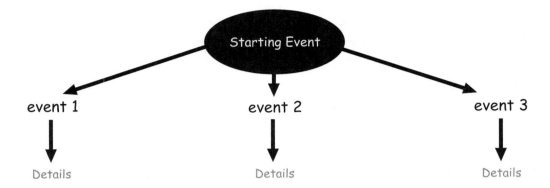

Starting Event

event 1   event 2   event 3

Details   Details   Details

or
_____

Event: _____

1. _____
   then
2. _____
   then
3. _____
   then
4. _____
   then
5. _____

**Series of Events Chain
or Flow Chart uses:** Description of stages, steps, sequences, or actions.

Examples include: The formation of something like clouds; the steps in a procedure or experimental design; a sequence of events like erosion or weathering; or the actions leading to the research and development of a product like television or a procedure like space exploration.

Questions to ask: How does the event or process begin? What are the next stages or steps? How are they connected? What is the end product or result?

Timeline

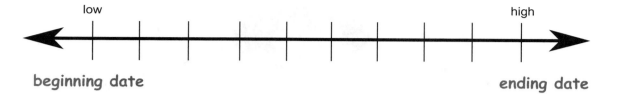

low   high

beginning date   ending date

Timeline uses: Showing historical events, creating a record, or charting events.

Examples include: The development of the light bulb, eras of geological time, a growth chart for plants, phases of the moon.

Questions to ask: When does the activity or process begin? When does it end? What happened on a specific date in time?

Venn Diagram

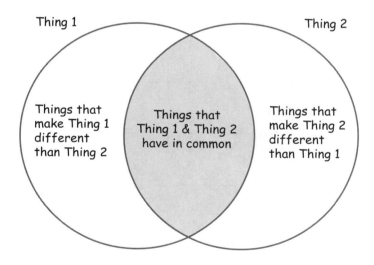

Venn Diagram uses: The comparison and contrast of objects, events, organisms, or themes.

Examples include: The comparison of the microscope and telescope. They both assist the eye in seeing objects. The microscope enlarges smaller objects; the telescope makes objects that are far away closer for observation.

Questions to ask: How are the objects, events, or themes the same? How are they different?

Chapter 1

The activities in this section of the book will focus on Science as Inquiry.

These activities will help you develop the skills necessary to do scientific inquiry and understand scientific inquiry. These activities include:

- identify questions that can be answered through scientific investigations.

- design and conduct a scientific investigation,

- use appropriate tools and techniques to gather, analyze, and interpret data,

- develop descriptions, explanations, predictions, and models using evidence,

- think critically and logically to make the relationships between evidence and explanations,

- recognize and analyze alternative explanations and predictions,

- communicate scientific procedures and explanations, and

- use mathematics in all aspects of scientific inquiry.

Use the "Clues for Success" Checklists as you complete each activity in this section as a tool to help you do your best work.

Step 1

Read the scenario "To Pop or Not to Pop."

To Pop or Not to Pop

The fifth grade science class had a popcorn party. They made the popcorn in the microwave oven. Each team received their own bag to pop. The bag had the following instructions. "Pop for 3–5 minutes or until you do not hear any more popping." Mary's group decided to time their bag for 3 minutes. Joel's group kept their bag in the microwave for the middle range time of 4 minutes. Tamara's group kept their bag in the longest recommended time of 5 minutes. And Vincent's group waited until they heard no more popping in the microwave oven, which was at least two minutes over the recommended time.

Miss Smith decided to use the results of the popped corn as a class experiment to see which time produced the best results for the popcorn bag. The class decided that they would use the amount of the popcorn produced as a measure of success.

They looked in the lab for tools they could use to measure their popcorn. They thought they might be able to use the scale or a plastic liter container.

The students did not like the plan of using either of these lab tools. They wanted to eat the popcorn! So they took a vote and decided they would describe the popcorn, first. Then, they would eat it as their snack. And finally, they would count the unpopped kernels to determine the best popping time.

They convinced their teacher because they said they would collect two types of data. One was descriptive (qualitative) and the other was numerical (quantitative). Here is the data each group posted on the chalkboard.

Group	Unpopped Kernels	Description of the Popcorn
Mary	40 unpopped	fluffy, white popcorn
Joel	70 unpopped	fluffy, white popcorn
Vincent	20 unpopped	clumpy and burnt
Tamara	30 unpopped	fluffy, white popcorn

Activity 1

Step 2

Complete the Checklist "Clues for Success."
The checklist will help you to read and think like a scientist.

Clues for Success

☐ **C**arefully read the information.

☐ **L**ook at any illustrations or diagrams.
They may provide you with additional information to answer the question.

☐ **U**nderstand the way you are asked to answer the question.
- ☐ Graph
- ☐ Chart
- ☐ Diagram
- ☐ Complete sentences
- ☐ Phrase
- ☐ Filled circle

☐ **E**xamine the information given.
- ☐ Reread the questions.
- ☐ Underline key words or phrases.
- ☐ Think about what the questions are asking.

☐ **S**ee if your answers match the questions.

Activity 1

Step 3

Use the information from
"To Pop or Not to Pop"
to complete the graphic organizer.

Make a chart to show the time each team used to pop their bag of popcorn.

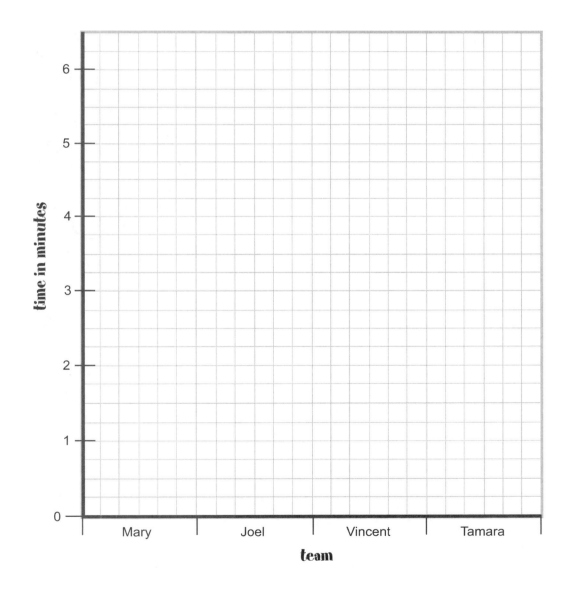

Activity 1

Step 4

Answer the following questions for "To Pop or Not to Pop" using information from your graphic organizer.

1. Select **one** of the science tools shown below. Then, list the steps students could use to measure the amount of their popcorn. Use complete sentences.

Liter Container

Scale

_____

_____

_____

_____

_____

_____

2. Create a graph that shows the number of unpopped kernels for each time the corn is popped. Be sure to title your graph.

title: _____

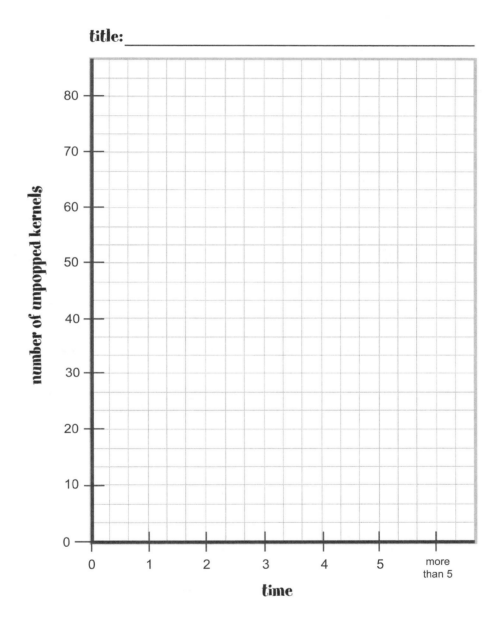

3. Which time was the best time for making fluffy, white popcorn with the least amount of unpopped kernels? Use a completely filled circle to show your answer choice.

    ○   3 minutes

    ○   4 minutes

    ○   5 minutes

    ○   until the corn stops popping

Use complete sentences to give evidence from the qualitative and quantitative data to explain your answer choice.

qualitative (descriptive) _____

_____

_____

_____

quantitative (numerical) _____

_____

_____

_____

Activity 2

Step 1

Read the scenario "Setting Up the Laboratory."

Setting Up the Laboratory

The students were setting up their lab tables for their plant experiment and observation. The teacher wanted them to be organized, so she made a step-by-step plan to help them set up their lab space.

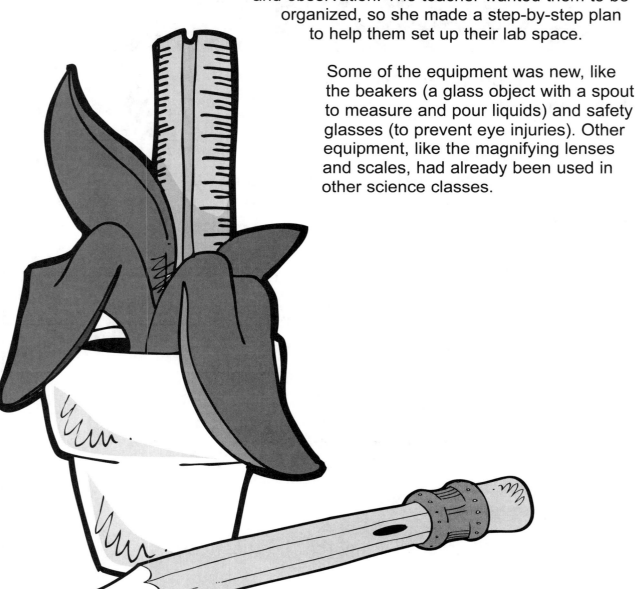

Some of the equipment was new, like the beakers (a glass object with a spout to measure and pour liquids) and safety glasses (to prevent eye injuries). Other equipment, like the magnifying lenses and scales, had already been used in other science classes.

Step 2

Complete the Checklist "Clues for Success."

The checklist will help you to read and think like a scientist.

Clues for Success

☐ **C**arefully read the information.

☐ **L**ook at any illustrations or diagrams.
They may provide you with additional information to answer the question.

☐ **U**nderstand the way you are asked to answer the question.
   ☐ Graph
   ☐ Chart
   ☐ Diagram
   ☐ Complete sentences
   ☐ Phrase
   ☐ Filled circle

☐ **E**xamine the information given.
   ☐ Reread the questions.
   ☐ Underline key words or phrases.
   ☐ Think about what the questions are asking.

☐ **S**ee if your answers match the questions.

Activity 2

Use the information from "Setting Up the Laboratory" to complete the graphic organizer.

Look at the illustrations of the objects on page 13. Then, read the sentences describing where the object should be placed on the lab table. Cut and glue the objects onto the lab table on page 12 using the directional terms provided.

a. The safety glasses are on top of the journal.
b. Place the scale on the top left corner of the lab table.
c. The markers are next to the plant, below the beaker.
d. The potted plant with ruler is in the middle of the table.
e. The thermometer is in the beaker of water.
f. The pencil is to the right of the journal.
g. The magnifying lens is on the bottom right of the lab table.
h. The notebook is on the bottom left corner of the lab table.
i. The beaker is on the top right corner of the lab table.

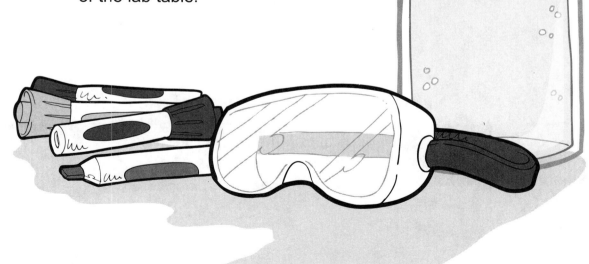

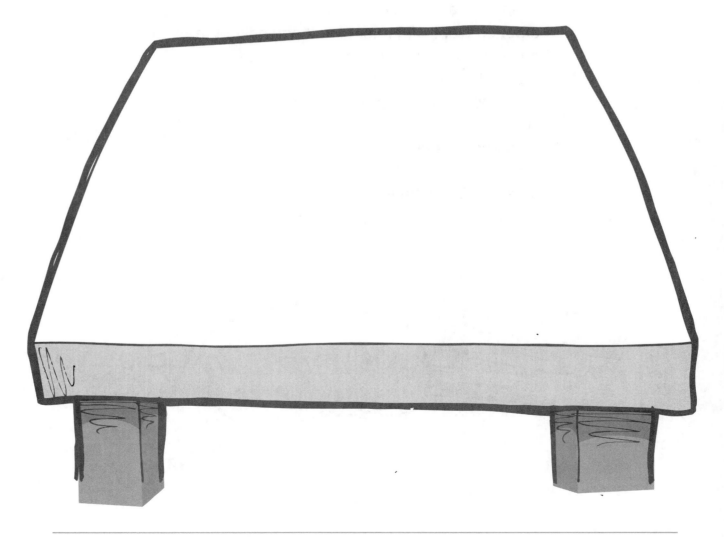

 © Englefield & Associates, Inc.

Step 4

Answer the following questions for "Setting Up the Laboratory" using information from your graphic organizer.

Read the data needed, then look at the lab table constructed in the scenario. Decide if all of the scientific equipment needed is available on the lab table to collect the named data. If you need additional material, write what is needed in the space provided

1. The difference in the height of two plants.

    ○ Available on the lab table   ○ Not available on the lab table

    Equipment needed:

    _____

    _____

    _____

2. The temperature of the room.

    ○ Available on the lab table   ○ Not available on the lab table

    Equipment needed:

    _____

    _____

    _____

3. The amount of water in the glass beaker.

   ○ Available on the lab table   ○ Not available on the lab table

   Equipment needed:

   _____

   _____

   _____

   _____

4. The number of days the plant has been growing.

   ○ Available on the lab table   ○ Not available on the lab table

   Equipment needed:

   _____

   _____

   _____

   _____

Step 1

Read the scenario
"Experimenting and Investigating."

Experimenting and Investigating

The fifth grade students were learning to design and conduct investigations. They learned they should:

1. Formulate a question,

2. design an investigation,

3. conduct an investigation,

4. interpret data,

5. use evidence to generate explanations,

6. propose alternative explanations, and

7. critique explanations and procedures.

Max's group wrote out these steps to their experimental design.

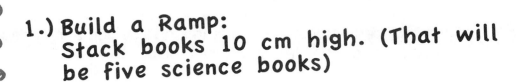

1.) Build a Ramp:
Stack books 10 cm high. (That will be five science books)

Use an encyclopedia volume as a ramp for the experiment.

Choose the surface for the car experiment.

Cover the surface of the experiment with large bulletin board paper.

2.) Select a car:
Dip the wheels in paint.

Let the car slide down the ramp.

Present the data to the class.

Step 2

Complete the Checklist "Clues for Success."

The checklist will help you to read and think like a scientist.

Clues for Success

☐ **C**arefully read the information.

☐ **L**ook at any illustrations or diagrams.
     They may provide you with additional information to answer the question.

☐ **U**nderstand the way you are asked to answer the question.
     ☐ Graph
     ☐ Chart
     ☐ Diagram
     ☐ Complete sentences
     ☐ Phrase
     ☐ Filled circle

☐ **E**xamine the information given.
     ☐ Reread the questions.
     ☐ Underline key words or phrases.
     ☐ Think about what the questions are asking.

☐ **S**ee if your answers match the questions.

Step 3

Use the information from "Experimenting and Investigating" to complete the graphic organizer.

Consider the students' scientific investigation, then draw a diagram to illustrate the group's experimental design.

                           © Englefield & Associates, Inc.

What do you think the group is investigating with this experimental design?

_____

_____

_____

_____

What information from the design leads you to this conclusion?

_____

_____

_____

_____

Activity 3

Step 4

Answer the following questions for "Experimenting and Investigating" using information from your graphic organizer.

Max's group had four trial runs using the car and the ramp. What might you conclude about each trial run from the paint track's data shown?

1. Trial I Data

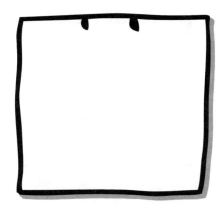

Conclusion: _____

Reason: _____

_____

2. Trial II Data

Conclusion: _____

Reason: _____

_____

3. Trial III Data

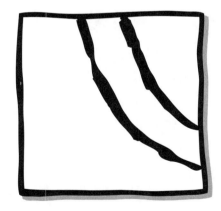

Conclusion: _____

Reason: _____

_____

4. Trial IV Data

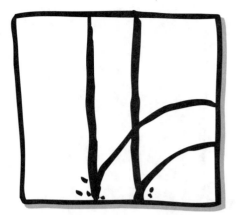

Conclusion: _____

Reason: _____

_____

Step 1

Read the scenario "The Science and Math Connection."

The Science and Math Connection

Bradley loved science, but he did not like math. He didn't see why or how math could help him be an investigator. Mr. Kramer, his math teacher, set out to teach Bradley the importance of mathematics by setting up the following scenarios. Mr. Kramer then challenged Bradley to investigate these situations and present the evidence (collect, display, and analyze) he found. Mr. Kramer told Bradley he could use any equipment from the science shelf, a calculator, and the classroom computer.

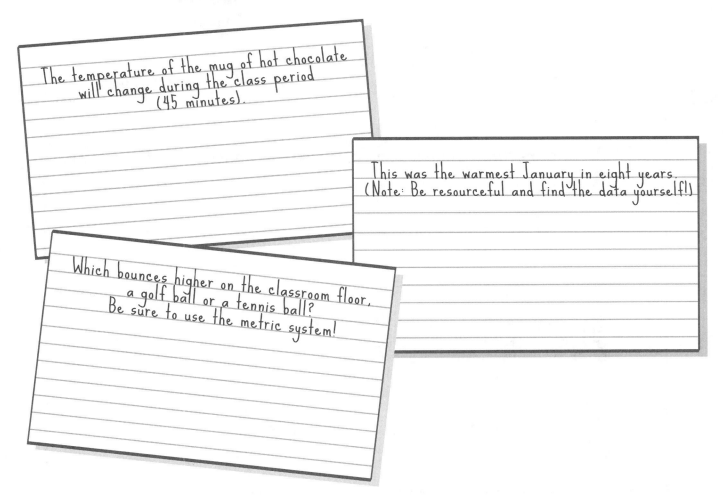

The temperature of the mug of hot chocolate will change during the class period (45 minutes).

This was the warmest January in eight years. (Note: Be resourceful and find the data yourself!)

Which bounces higher on the classroom floor, a golf ball or a tennis ball? Be sure to use the metric system!

Step 2

Complete the Checklist "Clues for Success."

The checklist will help you to read and think like a scientist.

Clues for Success

☐ **C**arefully read the information.

☐ **L**ook at any illustrations or diagrams.
    They may provide you with additional information to answer the question.

☐ **U**nderstand the way you are asked to answer the question.
    ☐ Graph
    ☐ Chart
    ☐ Diagram
    ☐ Complete sentences
    ☐ Phrase
    ☐ Filled circle

☐ **E**xamine the information given.
    ☐ Reread the questions.
    ☐ Underline key words or phrases.
    ☐ Think about what the questions are asking.

☐ **S**ee if your answers match the questions.

Step 3

Use the information from "The Science and Math Connection" to complete the graphic organizer.

Complete the chart below to indicate the scientific tools Bradley used to collect the data and complete his investigation. In the tool column, be sure to tell which tool he used to perform the mathematical calculations. Then, tell the mathematical processes Bradley used to analyze his data.

	Scientific Tools Used	Mathematical Processes
The temperature of the hot chocolate changed		
This was the warmest January in eight years		
Which was the higher bouncing ball		

Step 4

Answer the following questions for "The Science and Math Connection" using information from your graphic organizer.

1. Bradley used a graph to display the data for the change in temperature of the mug of hot chocolate. What will the graph look like?

On the graph below, label the *x*- and *y*- axes, and add the title of the graph.

title: _____

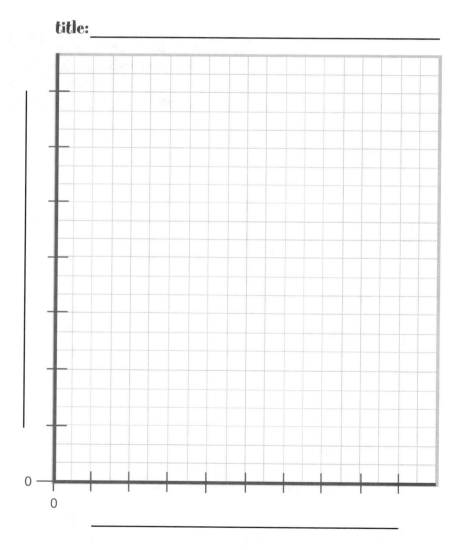

2. Bradley used a bar graph to present the data he found for the last eight years.

Average January Temperatures								
Year:	1	2	3	4	5	6	7	8
Average Temp.	21	25	23	30	29	22	23	32

Create the temperature bar graph Bradley presented to Mr. Kramer.

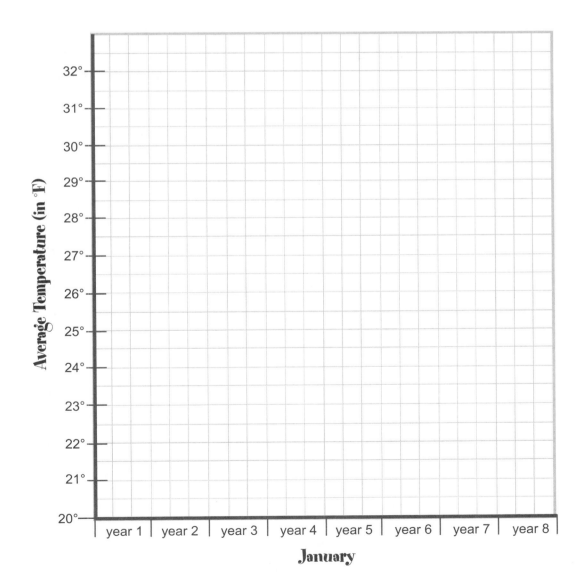

3. Bradley bounced each of the balls 5 times and recorded the results.

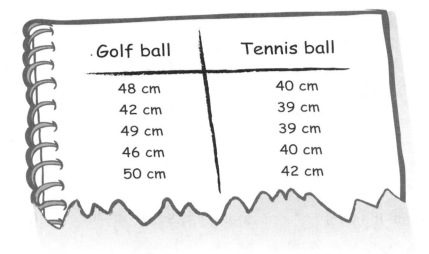

Golf ball	Tennis ball
48 cm	40 cm
42 cm	39 cm
49 cm	39 cm
46 cm	40 cm
50 cm	42 cm

Find the average height for each ball bounced, label and circle the average heights on the index card, and draw a star by the highest average.

Golf Ball	Tennis Ball

Chapter 2

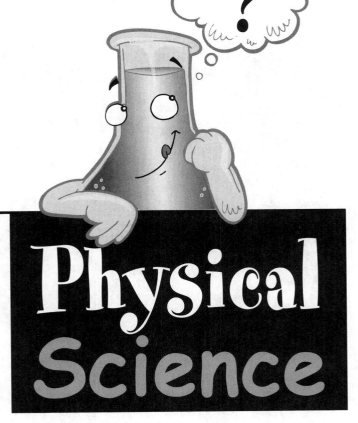

Physical Science

The activities in this section of the book will focus on Physical Science.

These activities will help you investigate ideas about objects and materials that include:

- properties and changes of properties in matter,
- motions and forces, and
- transfer of energy.

Use the "Clues for Success" Checklists as you complete each activity in this section as a tool to help you do your best work.

Step

1

Read the scenario "State of Matter."

State of Matter

The students were learning about the states of matter. They learned that matter can exist in different states: solid, liquid, and gas.

They wrote what they learned on the chalkboard.

	description
solids	molecules in slight motion, fixed in position have a definite shape
liquids	molecules in motion, take the shape of their container
gas	molecules in rapid motion take the shape of their container

States of matter may be changed by increasing the energy of the molecules or decreasing the energy of the molecules.

Step 2

Complete the Checklist "Clues for Success."

The checklist will help you to read and think like a scientist.

Clues for Success

☐ **C**arefully read the information.

☐ **L**ook at any illustrations or diagrams.
They may provide you with additional information to answer the question.

☐ **U**nderstand the way you are asked to answer the question.
　　☐ Graph
　　☐ Chart
　　☐ Diagram
　　☐ Complete sentences
　　☐ Phrase
　　☐ Filled circle

☐ **E**xamine the information given.
　　☐ Reread the questions.
　　☐ Underline key words or phrases.
　　☐ Think about what the questions are asking.

☐ **S**ee if your answers match the questions.

Step 3

Use the information from
"State of Matter"
to complete the graphic organizer.

Look at each object below. Follow the arrow to its new state of matter.
Label if the molecules "increase" or "decrease" motion (movement) as
they change into this new state.

solid

molecules

_____

motion

molecules

_____

motion

liquid

molecules

_____

motion

molecules

_____

motion

gas

                   © Englefield & Associates, Inc.

Activity 1

Step 4

Answer the following questions for "State of Matter" using information from your graphic organizer.

1. Write on the arrow the change in state of matter on the following substances. Then, on the line provided, give a possible source of the energy change.

Source of energy: _____

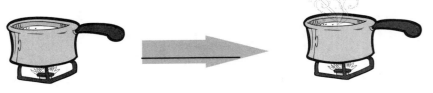

Source of energy: _____

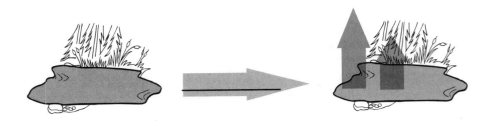

Source of energy: _____

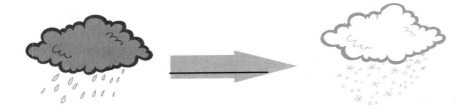

Source of energy: _____

2. Describe a series of events that would affect the gain or loss of energy in the molecules of a crayon left in a car on a hot summer's day until it was found the next morning. Be sure to include what the crayon physically looked like at the beginning of the day and what the crayon looked like when it was found the next morning.

Describe the events affecting the gain or loss of energy in the crayon.

_____

_____

_____

_____

_____

_____

Activity 2

Step 1

Read the scenario "Properties of Matter."

Properties of Matter

The students in the fifth grade were learning about substances. They learned that every substance has specific physical properties. These properties stay the same no matter how much of that particular substance is available. Color and hardness of a substance are examples of these qualities that can easily be **observed**.

Some other physical properties must be **measured** with scientific tools. These characteristics include size, the ability to conduct electricity, the boiling and freezing points, and the ability to be dissolved in other substances. If there is a mixture of substances, they can be separated by using these characteristics.

Some physical properties can be determined combining several measurements. One of these properties is density. Density can be determined by dividing the mass of the object (amount of matter) by the volume (the amount of space the object takes).

Step 2

Complete the Checklist "Clues for Success."

The checklist will help you to read and think like a scientist.

Clues for Success

☐ **C**arefully read the information.

☐ **L**ook at any illustrations or diagrams.
They may provide you with additional information to answer the question.

☐ **U**nderstand the way you are asked to answer the question.
☐ Graph
☐ Chart
☐ Diagram
☐ Complete sentences
☐ Phrase
☐ Filled circle

☐ **E**xamine the information given.
☐ Reread the questions.
☐ Underline key words or phrases.
☐ Think about what the questions are asking.

☐ **S**ee if your answers match the questions.

Step 3

Use the information from "Properties of Matter" to complete the graphic organizer.

Complete the flow chart to organize the information about physical properties of matter.

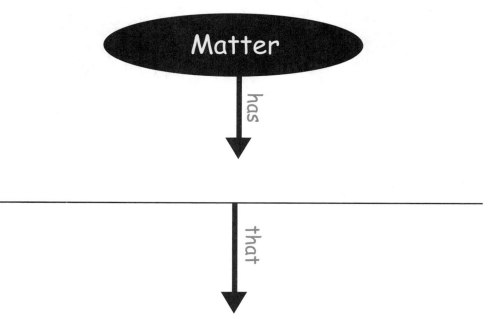

Matter

has

that

Stay the same no matter how much of that substance is available.

They can be.

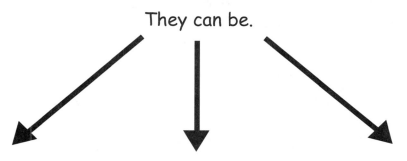

_____   _____   _____

Activity 2

Step 4

Answer the following questions for "Properties of Matter" using information from your graphic organizer.

The students were given three cubes (wood, copper, and aluminum) to investigate and describe using physical characteristics.

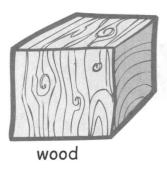
wood

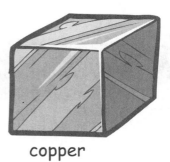
copper

aluminum

1. Hava said the blocks were 2 centimeter squares. Which physical property did she describe?

   _____

2. Chad said the blocks looked different: two were shiny, one was golden, and one was silvery. He said one looked dull and beige with dark lines. Which physical property did he describe?

   _____

3. Shana arranged them in a line according to their mass from least to greatest: wood, aluminum, and copper. How would Shana get that information?

   _____

4. YK wanted to measure the density of the blocks. He said he could use the measurements of the other students in his group. Which students' measurements will he use?

   ○ Hava's   ○ Chad's   ○ Shana's

   Why would he use those measurements?

   _____

5. Max was playing around with the cubes. He dropped the cubes in a beaker of water. The copper cube dropped to the bottom, and the aluminum and wood cubes floated at the top.

   What did Max observe about the density of the blocks?

   _____

   _____

   _____

6. Then Max crumbled some aluminum foil into balls about 1 centimeter in diameter and some balls about 3 centimeters in diameter and dropped them into the water. Both sized aluminum balls floated like the aluminum cube.

   What did Max observe about the physical properties of aluminum?

   _____

   _____

   _____

Step 1

Read the scenario
"Motions and Forces."

Motions and Forces

Josh took these notes about motions and forces when he read his homework assignment.

Forces interact on earth to create motion. The motion of an object can be described by its position, direction of motion, and speed (units of distance divided by units of time).

Motion can be measured and illustrated to show the change in position, direction, and speed.

If an object does not experience a force, it will remain at rest (stopped), or if it is moving it will continue to move at a constant speed and in a straight line.

Forces that act on an object include pushes or pulls that cause an object to move, stop, change speed, or change directions.

If more than one force acts on an object along a straight line, then the forces will reinforce or cancel one another, depending on their direction and magnitude.

Unbalanced forces will cause changes in the speed or direction of an object's motion.

Step 2

Complete the Checklist "Clues for Success."
The checklist will help you to read and think like a scientist.

Clues for Success

☐ **C**arefully read the information.

☐ **L**ook at any illustrations or diagrams.
   They may provide you with additional information to answer the question.

☐ **U**nderstand the way you are asked to answer the question.
   ☐ Graph
   ☐ Chart
   ☐ Diagram
   ☐ Complete sentences
   ☐ Phrase
   ☐ Filled circle

☐ **E**xamine the information given.
   ☐ Reread the questions.
   ☐ Underline key words or phrases.
   ☐ Think about what the questions are asking.

☐ **S**ee if your answers match the questions.

Step 3

Use the information from "Motions and Forces" to complete the graphic organizer.

To help them study for a quiz on forces, the students received this study guide. Answer each question using Josh's class notes.

STUDY GUIDE

What three things can be used to describe the motion of an object?

1. _____

2. _____

3. _____

What will happen to an object if it does not experience a force?

1. _____

2. _____

 　© Englefield & Associates, Inc.

Using the information from Josh's notes, draw either a ←, →, or —
under the illustration to show what would happen if two pennies moving
with the same force directly hit each other.

What information from Josh's notes did you use?

_____

_____

_____

Using the information from Josh's notes, draw an arrow under the
illustration to show what would happen if a quarter moving with the
same force directly hit a penny.

What information from Josh's notes did you use?

_____

_____

_____

Activity 3

Step 4

Answer the following questions for "Motions and Forces" using information from your graphic organizer.

George brought his skate board to school to illustrate motion to the class. The class went out to the playground. The students were given the following tasks:

Illustrate George's motion.

Indicate his beginning and ending position with a star.

Use arrows to indicate the direction of his motion.

George began and ended his demonstration at the same point near the swing set. He kept a constant motion as he traveled the perimeter of the playground. His time was clocked at 180 seconds.

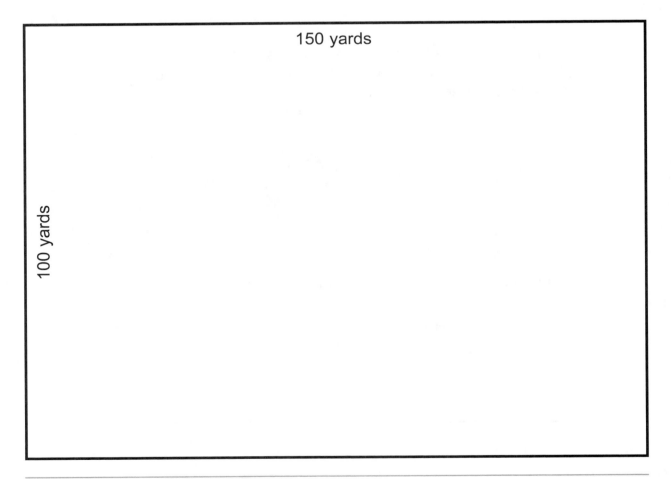

1. How far did George travel in the 180 seconds?

_____

While they were at the playground the students used the teeter-totter to demonstrate balanced and unbalanced forces.

2. Draw a teeter-totter, showing **balanced** forces.

Explain your illustration.

_____

_____

_____

_____

3. Draw a teetor-totter, showing **unbalanced** forces.

Explain your illustration.

_____

_____

_____

_____

Activity 4

Step 1

Read the scenario "Sun Energy."

Sun Energy

The sun is the source of energy on Earth. As the sun loses its energy, a small portion of it travels to Earth in the form of light energy. The light is energy in wave form called electromagnetic radiation. The longest waves have the least energy; the shortest waves have the most energy. Although there are no clear separations between categories, several ranges of the wavelengths characteristics have been described. They include infrared radiation, visible light, ultraviolet radiation, and X-rays.

Infrared radiation has a long wavelength and low energy. Electromagnetic radiation is felt as heat. The lower part of infrared radiation may be called microwaves. Objects give off heat in this range and can give an idea of what kind of activity is occurring in a specific area. For example, home owners can get an idea where heat is escaping from their homes using infrared screenings.

Visible light is the part of the sun's radiations that we can see. Visible light is composed of a various wavelengths. A range of wavelengths (colors) can be seen as visible light passes through a prism. From the longest wavelength, to the shortest wavelength the color range is red (longest), orange, yellow, green, blue, indigo, and violet (shortest). Red has the least energy, and violet has the most energy.

Ultraviolet (UV) radiation is a type of electromagnetic radiation with very short wavelengths below those of the color blue in visible light. Ultraviolet rays are invisible to us. Because UV rays are high in energy, they can cause changes in substances. UV radiation can change skin cells in the human body causing sunburn or skin cancer. The sun releases a large amount of UV radiation with most of it absorbed by the ozone layer in the atmosphere before reaching Earth.

X-rays have a shorter wavelength and more energy than UV radiation. X-rays are useful in medicine, industry, and space exploration because they can pass through most substances.

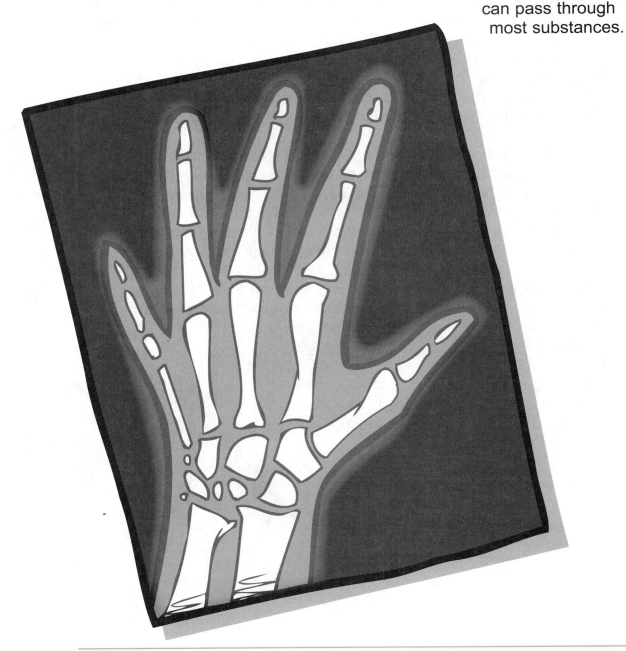

Step 2

Complete the Checklist "Clues for Success."

The checklist will help you to read and think like a scientist.

Clues for Success

☐ **C**arefully read the information.

☐ **L**ook at any illustrations or diagrams.
  They may provide you with additional information to answer the question.

☐ **U**nderstand the way you are asked to answer the question.
  ☐ Graph
  ☐ Chart
  ☐ Diagram
  ☐ Complete sentences
  ☐ Phrase
  ☐ Filled circle

☐ **E**xamine the information given.
  ☐ Reread the questions.
  ☐ Underline key words or phrases.
  ☐ Think about what the questions are asking.

☐ **S**ee if your answers match the questions.

Step 3

Use the information from "Sun Energy" to complete the graphic organizer.

Complete the table, listing **two** characteristics about each type of radiation.

	Characteristics
Infrared radiation	1. _____ _____ 2. _____ _____
Visible light	1. _____ _____ 2. _____ _____
Ultraviolet radiation	1. _____ _____ 2. _____ _____
X-rays	1. _____ _____ 2. _____ _____

 © Englefield & Associates, Inc.

Activity 4

Step 4

Answer the following questions for "Sun Energy" using information from your graphic organizer.

1. Place X-rays, UV Radiation, Visible Light, and Infrared Radiation on the chart below. Complete the blanks to describe the energy of each wavelength.

Longest wavelength   Shortest wavelength

_____ energy   _____ energy

2. Why is the ozone layer important?

_____

_____

3. How might X-rays be used if someone falls off of their bicycle?

_____

_____

© Englefield & Associates, Inc.

Chapter 3

The activities in this section of the book will focus on Life Science.

These activities will help you investigate:

- structure and function in living systems,

- reproduction and heredity,

- regulation and behavior,

- populations and ecosystems, and

- diversity and adaptations of organisms.

Use the "Clues for Success" Checklists as you complete each activity in this section as a tool to help you do your best work.

Step 1

Read the scenario "Ecosystem Reports."

Ecosystem Reports

The class was studying how the needs of living things are met in different environments called ecosystems. The ecosystem provides the air, water, food, nutrients, and light the organisms need to live. The students researched and presented reports on various ecosystems. Read the reports about the rainforest and coral reef ecosystems below. Use the information to help you compare these two ecosystems.

```
                    Rainforests
                    by Jerome


    A rainforest is a land environment. Many
rainforests are found near the equator. They get a
lot of rainfall and have high temperatures all year
long. The rainforest has a great variety (diversity)
of plants and animals. The different kinds of
organisms that live there include mammals (monkeys),
birds (toucans), insects (morpho butterflies), trees
(kapok), fish (piranhas), and reptiles (anacondas).
These organisms live in different parts of the
rainforest and receive different amounts light
because some of them need shade and others need
sunshine. There are several life zones in the
rainforest. These layers of life include the canopy
(upper level), middle level, and the forest floor.
The rotting plants on the forest floor (decomposing)
recycle nutrients in the ecosystem.
```

A Coral Reef

by Gracie

If you want to find a coral reef, you should look along the equator. Coral reefs are generally found in a tropical climate and underwater. The water must have a specific amount of salt. This amount is important and it is called salinity.

All of the areas of a coral reef do not get the same amount of sunshine. Some areas of the reef receive light and some are shady.

A coral reef is made up of skeletons of animals called corals. When these animals die, other coral polyps move into the skeletons they leave behind. Coral reefs also have branches. These coral skeletons have holes and cracks where other creatures like crabs, octopi, anemone, and fish can live.

Step 2

Complete the Checklist "Clues for Success."
The checklist will help you to read and think like a scientist.

Clues for Success

☐ **C**arefully read the information.

☐ **L**ook at any illustrations or diagrams.
    They may provide you with additional information to answer the question.

☐ **U**nderstand the way you are asked to answer the question.
    ☐ Graph
    ☐ Chart
    ☐ Diagram
    ☐ Complete sentences
    ☐ Phrase
    ☐ Filled circle

☐ **E**xamine the information given.
    ☐ Reread the questions.
    ☐ Underline key words or phrases.
    ☐ Think about what the questions are asking.

☐ **S**ee if your answers match the questions.

Step

3

Use the information from "Ecosystem Reports" to complete the graphic organizer.

Summarize the following information from "Ecosystem Reports."

STUDY GUIDE

How are the rainforest and coral reefs the same?
Give <u>two</u> examples.

1. _____

   _____

2. _____

   _____

How are the rainforest and coral reefs different?
Give <u>two</u> examples.

1. _____

   _____

2. _____

   _____

Create a Venn Diagram to compare and contrast the rainforest and the coral reef. Be sure to label the parts of the Venn Diagram.

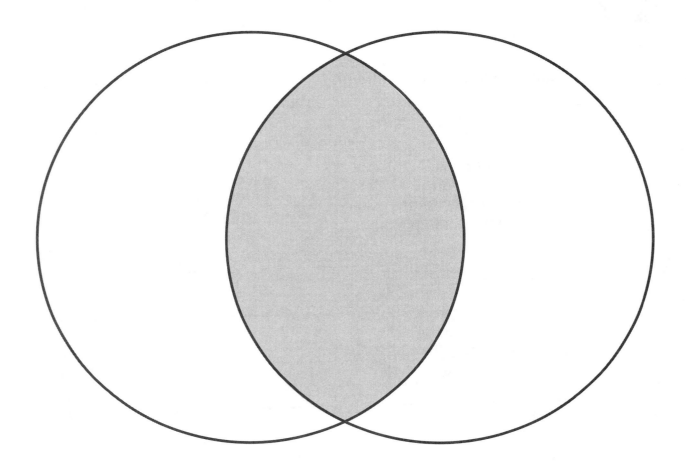

Step 4

Answer the following questions for "Ecosystem Reports" using information from your graphic organizer.

1. Using the information about the rainforest and the coral reef, do you think an organism could live in both of the ecosystems? Use a completely filled circle to show your answer choice.

    ○ Yes   ○ No

    Give examples to support your answer.

    _____

    _____

    _____

    _____

    _____

    _____

    _____

Activity 2

Step 1

Read the scenario
"Organisms and Their Environments."

Organisms and Their Environments

Organisms create changes in their environment (the place where they live).

Some of these changes help or benefit the environment. Some examples of these interactions include:
- Bacteria decomposing material and
- People cleaning up litter.

Other changes create problems or harm the environments. Some examples of these interactions include:
- Unwanted plants (weeds) growing in the lawn and
- Tree roots cracking the sidewalk.

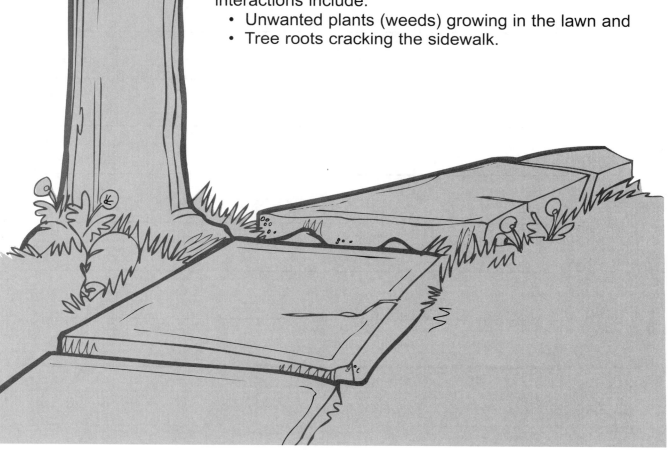

Step 2

Complete the Checklist "Clues for Success."

The checklist will help you to read and think like a scientist.

Clues for Success

☐ **C**arefully read the information.

☐ **L**ook at any illustrations or diagrams.
　　They may provide you with additional information to answer the question.

☐ **U**nderstand the way you are asked to answer the question.
　　☐ Graph
　　☐ Chart
　　☐ Diagram
　　☐ Complete sentences
　　☐ Phrase
　　☐ Filled circle

☐ **E**xamine the information given.
　　☐ Reread the questions.
　　☐ Underline key words or phrases.
　　☐ Think about what the questions are asking.

☐ **S**ee if your answers match the questions.

Step 3

Use the information from "Organisms and Their Environments" to complete the graphic organizer.

Read each statement on page 65. Each statement is about a change an organism creates in its environment. Decide if the change **helps** or **harms** the environment. Put an X on the actions that hurt the environment. Cut out the statements, sort them into good (helpful) things or problems (harmful). Finally, glue the statements onto the appropriate web.

helps the environment	harms the environment

Beavers build a dam that flood a farmer's field.

The trees roots are coming up through the sidewalk.

A gypsy moths eat so many leaves that a tree dies.

People are cleaning up the stream.

Ladybugs eat the insects that are destroying the farmer's crops.

Trees provide a place for birds to build nests.

There is a compost pile to help fertilize the garden.

The fifth graders are planting a vegetable garden.

Weeds are growing in the garden.

The deer are eating all the vegetation in the park.

The moles are digging holes in the field.

© Englefield & Associates, Inc.

Step 4

Answer the following questions for "Organisms and Their Environments" using information from your graphic organizer.

1. One organism can be found in both sections "Helps the Environment" and "Harms the Environment." Which organism is it?

_____

2. Explain how this organism can both help and harm the environment.

_____

_____

_____

_____

_____

_____

_____

The fifth graders are planting a vegetable garden. List **two** examples of how the garden will create changes in the environment. Use a completely filled circle to show if the effect is positive (helpful) or negative (harmful). On the lines below, give a reason for your answer.

3. _____

○ Positive (helpful)   ○ Negative (harmful)

Reason:

_____

_____

_____

_____

4. _____

○ Positive (helpful)   ○ Negative (harmful)

Reason:

_____

_____

_____

_____

Step 1

Read the scenario
"Flowering Plants and Sexual Reproduction."

Flowering Plants and Sexual Reproduction

The class was studying plants. They were surprised to learn that some plants reproduced sexually. They learned that flowering plants called angiosperms produce seeds in a fruit. The fruit is a part of the reproductive system of the plant called the ovary. The ovary is a part of the female reproductive organs in a plant.

For a plant to reproduce sexually, the females produce eggs and males produce pollen that unite and form a seed. This process is called pollination. Animals, insects, and birds help in this process. Bees are common pollinators. A seed has the ability grow into a new individual. This individual has similar characteristics to it parents but is not identical to them. The process of reproduction is necessary for the organisms to continue on.

The female parts of the plant are collectively referred to as the pistol or carpel. They include:

- A stigma which receives the pollen during fertilization
- The style which is a tube at the top of the ovary. It is connected to the stigma.
- The ovary which is the place where the ovules (eggs) are produced.
- The ovules which are the female reproductive cells (eggs).

The male parts of the plant are collectively called the stamen. They include:

- The anther which produces the pollen, the male reproductive cell.
- The filament which is a thin stalk that holds the anther.

Flowering plants are classified as perfect or imperfect. A perfect flowering plant contains both male and female structures. Examples of perfect flowering plants include lilies, apples, and tomatoes. Imperfect flowers are either male or female. Examples of imperfect flowering plants include maize and squash. The tassel on top of a maize plant is male (and imperfect). Maize silks are female (and imperfect).

Step 2

Complete the Checklist "Clues for Success."

The checklist will help you to read and think like a scientist.

Clues for Success

☐ **C**arefully read the information.

☐ **L**ook at any illustrations or diagrams.
    They may provide you with additional information to answer the question.

☐ **U**nderstand the way you are asked to answer the question.
    ☐ Graph
    ☐ Chart
    ☐ Diagram
    ☐ Complete sentences
    ☐ Phrase
    ☐ Filled circle

☐ **E**xamine the information given.
    ☐ Reread the questions.
    ☐ Underline key words or phrases.
    ☐ Think about what the questions are asking.

☐ **S**ee if your answers match the questions.

Step 3

Use the information from "Flowering Plants and Sexual Reproduction" to complete the graphic organizer.

Label the male and female reproductive organs of the flowering plant. Then, use completely filled circles to label the collection of the male and the female parts beneath them on the diagram.

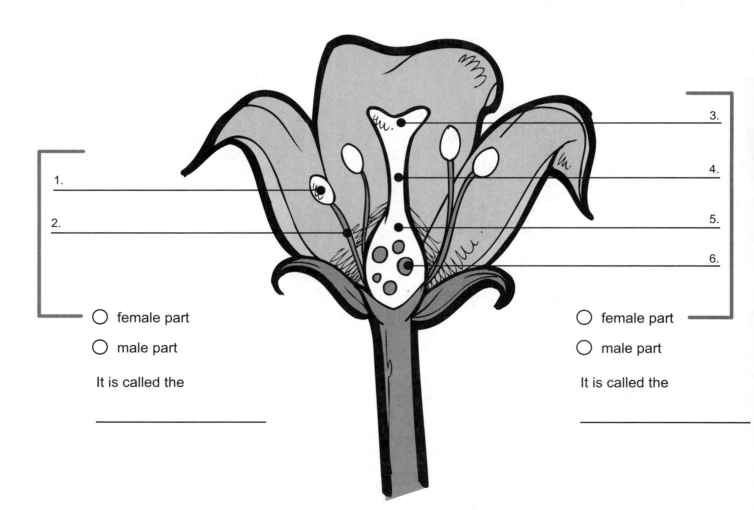

1.

2.

3.

4.

5.

6.

○ female part

○ male part

It is called the

_____

○ female part

○ male part

It is called the

_____

Activity 3

Step 4

Answer the following questions for
"Flowering Plants and Sexual Reproduction"
using information from your graphic organizer.

1. Look at the position of the stamen. How does its location on the flower help
   in the process of pollination of flowering plants?

   _____

   _____

   _____

   _____

2. Look at the position of the ovary in a flowering plant. How does its position
   help in plant reproduction?

   _____

   _____

   _____

   _____

3. Use your own words to describe the process of sexual reproduction in perfect flowering plants.

_____

_____

_____

_____

_____

_____

© Englefield & Associates, Inc.

Step 1

Read the scenario "To Stay or Not To Stay."

To Stay or Not To Stay

The science class was learning that organisms must be able to adapt to changes in their environment. Jennifer was going to the metro park for a birding work shop, so she asked the naturalist for more information about how birds adapt for the winter. Here is the information Jennifer shared with her classmates.

Birds respond to the seasonal changes in their environment by staying and responding to the changes or by migrating to a different environment.

If the birds stay, their most vital job is to stay warm and maintain their bodies' temperature. Birds puff out their feathers when it is cold to maintain a constant temperature. They also raise and lower their feathers to create insulation for their bodies. With this action, the birds create pockets of heat. Some birds even have down feathers. To conserve energy, the birds lower their body metabolism (functions like heart rate) and their body temperature at night. Birds might use bird houses or  natural shelters to stay warm. Some species like ducks have adaptations to help them in the colder weather. Ducks have a natural protection with the oil in their feathers that helps them swim in cold water. They also have a special circulatory mechanism that keeps pumping warm blood from the heart to the legs.

If the bird migrates to maintain life functions, there are several methods different species use. Some birds are born with an internal map of the stars and moon that helps them navigate during migrations. Some birds find their way using Earth's magnetic fields. The sound of the ocean waves help some birds migrate. Other species follow older birds in their population that have previously migrated to the new climate.

Scientists continue to study the ways birds migrate and adapt to the changing weather.

Step 2

Complete the Checklist "Clues for Success."

The checklist will help you to read and think like a scientist.

Clues for Success

☐ **C**arefully read the information.

☐ **L**ook at any illustrations or diagrams.
They may provide you with additional information to answer the question.

☐ **U**nderstand the way you are asked to answer the question.
☐ Graph
☐ Chart
☐ Diagram
☐ Complete sentences
☐ Phrase
☐ Filled circle

☐ **E**xamine the information given.
☐ Reread the questions.
☐ Underline key words or phrases.
☐ Think about what the questions are asking.

☐ **S**ee if your answers match the questions.

Step 3

Use the information from "To Stay or Not To Stay" to complete the graphic organizer.

The class took notes on Jennifer's presentation. List the details the students recorded for each of the topics.

Birds that migrate when the climate changes might use one of the following techniques:

1. _____

2. _____

3. _____

4. _____

Birds that stay in the area when the climate changes might use one of the following adaptations:

1. _____

2. _____

3. _____

4. _____

5. _____

Step 4

Answer the following questions for "To Stay or Not To Stay" using information from your graphic organizer.

1. How might having special oil in their feathers helps ducks swim in cold weather?

_____

_____

_____

2. Why might lowering a bird's body temperature at night help conserve energy?

_____

_____

_____

3. Why do you think it is important for scientists to keep studying about how birds migrate to different climates?

_____

_____

_____

Step 1

Read the scenario
"Skin—It's All Over You."

Skin—It's All Over You

Skin is the largest organ of the human body. The skin is built in the same manner as all of the other organs. Cells create tissues. Tissues create organs. And, organs create organisms.

Although it is about 2 mm thick, the skin performs several functions for the body. These functions include:
- Providing a barrier against germs,
- Keeping muscles, bones, and other internal organs inside the body,
- Maintaining a constant body temperature, and
- Providing the sense of feeling and touch.

The skin is composed of three different layers of tissues.

Epidermis is a tough protective layer made up of epithelial tissues. These tissues are tightly packed continuous sheets. These old, tough cell tissues protect the inside of the body. Epidermis cells wear off and are constantly being replaced.

Dermis contains nerve endings, oil glands, and hair follicles. The dermis also nourishes the skin cells with blood vessels that bring oxygen and nutrients to the cells and remove cellular waste. This layer is composed of connective tissue containing strands of collagen (a protein) that adds strength. The nerve tissues found in the dermis provide the sense of feeling and touch in the skin. Muscle tissues that contract and expand are found in the dermis. Hair and glands are also in this layer.

Subcutaneous layer is the fatty layer of the skin. It is located below the dermis. It is composed of adipose tissue that acts as padding for the skin. This layer helps the body stay warm in cold temperatures.

Step 2

Complete the Checklist "Clues for Success."

The checklist will help you to read and think like a scientist.

Clues for Success

☐ **C**arefully read the information.

☐ **L**ook at any illustrations or diagrams.
They may provide you with additional information to answer the question.

☐ **U**nderstand the way you are asked to answer the question.
   - ☐ Graph
   - ☐ Chart
   - ☐ Diagram
   - ☐ Complete sentences
   - ☐ Phrase
   - ☐ Filled circle

☐ **E**xamine the information given.
   - ☐ Reread the questions.
   - ☐ Underline key words or phrases.
   - ☐ Think about what the questions are asking.

☐ **S**ee if your answers match the questions.

Step 3

Use the information from
"Skin—It's All Over You"
to complete the graphic organizer.

EPIDERMIS

Type of Tissue   Function

_____   _____

DERMIS

Type of Tissues   Function

_____   _____

_____   _____

_____   _____

SUBCUTANEOUS LAYER

Type of Tissue   Function

_____   _____

Activity 5

Step 4

Answer the following questions for "Skin—It's All Over You" using information from your graphic organizer.

Consider each layer of the skin. In the spaces provided, use a sentence to describe a situation when the specific layer of the skin is important to the body.

1. Epidermis _____

_____

_____

2. Dermis _____

_____

_____

3. Subcutaneous layer _____

_____

_____

Chapter 4

The activities in this section of the book will focus on Earth Science.

These activities will focus on:

- structure of Earth system,

- Earth's history, and

- Earth in the solar system.

Use the "Clues for Success" Checklists as you complete each activity in this section as a tool to help you do your best work.

Step

1

Read the scenario
"What Goes Around Comes Around."

What Goes Around Comes Around

The objects in the solar system move in patterns that are regular and predictable. As an example to illustrate these patterns, the teacher offered several illustrations using the relationship of the sun, Earth, and moon in our solar system.

Most of the students had some knowledge about the relationship that creates the period of the day. The class discussed the rotation of Earth on its axis. The time when the sun is visible to parts of Earth is day (generally having light) and night (generally absence of light). The time period for a day is a period of about 24 hours. The students were also able to relate the motion that creates a year as the revolution of Earth on its orbit (path) around the sun. This time period is about 365.24 days.

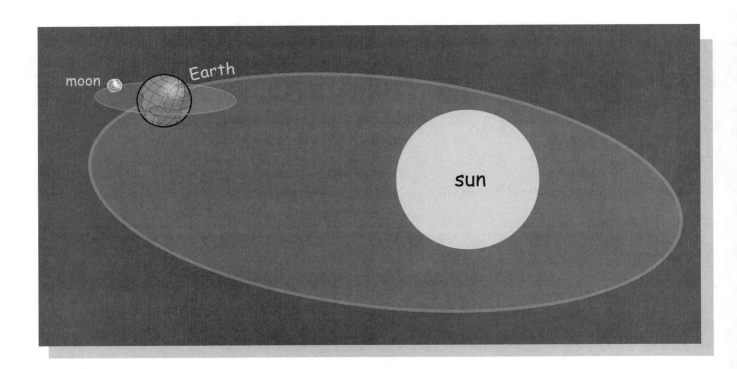

The students noted that the moon also moves in a predictable pattern about every 28 days. The moon does not have its own light; it is illuminated by the sun. The portion of the moon that can be seen depends on the location of the moon in its orbit around the sun. This pattern is called the phases of the moon. When the moon is completely illuminated by the sun, it is called a full moon. When half of the moon is illuminated by the sun, this is called the half moon. The new moon is not illuminated at all by the sun. Between these phases, crescent portions of the moon can be seen. They are crescent (between new and quarter moon) and gibbous (between quarter and full moon).

At the end of the lesson, the teacher hinted that even more regular and predictable patterns exist in the universe.

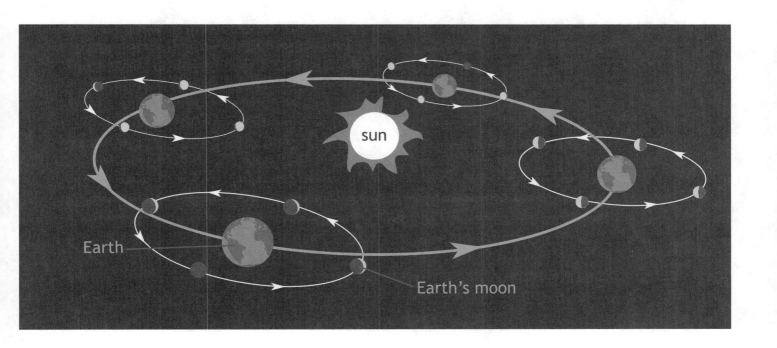

Step 2

Complete the Checklist "Clues for Success."

The checklist will help you to read and think like a scientist.

Clues for Success

- ☐ **C**arefully read the information.

- ☐ **L**ook at any illustrations or diagrams.
  They may provide you with additional information to answer the question.

- ☐ **U**nderstand the way you are asked to answer the question.
  - ☐ Graph
  - ☐ Chart
  - ☐ Diagram
  - ☐ Complete sentences
  - ☐ Phrase
  - ☐ Filled circle

- ☐ **E**xamine the information given.
  - ☐ Reread the questions.
  - ☐ Underline key words or phrases.
  - ☐ Think about what the questions are asking.

- ☐ **S**ee if your answers match the questions.

© Englefield & Associates, Inc.

Step 3

Use the information from "What Goes Around Comes Around" to complete the graphic organizer.

Use complete sentences to summarize the information from the scenario, "What Goes Around Comes Around," in the outline provided.

I. The main topic of the lesson

   A. Pattern that involves the movement of Earth

   B. Pattern that involves the movement of the moon

   C. Patterns that involve the sun

Step 4

Answer the following questions for "What Goes Around Comes Around" using information from your graphic organizer.

1. Draw and label the pattern that creates a day.

2. Draw and label the pattern that creates a year.

3. Draw and label the phases of the moon.

Step 1

Read the scenario "Earth's Atmosphere."

Earth's Atmosphere

Earth is surrounded by a thin layer of gases called the atmosphere.

The atmosphere was formed by processes that released gases from the interior of Earth and activities such as volcanic eruptions. The major gases that make up the atmosphere include nitrogen (71%) and oxygen (21%) and various other gases. The atmosphere is important because it helps keep Earth's temperature within the range for life. The atmosphere keeps the heat near Earth while keeping out the harmful rays of the sun's ultraviolet radiation.

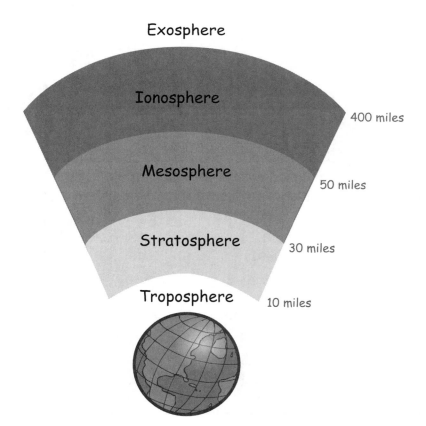

Most of Earth's atmosphere is 10 miles (16 kilometers) from the surface of Earth. There is no exact place where the atmosphere ends; it just gets thinner and thinner until it merges with the space surrounding it.

There are five layers of gases with different characteristics as they move out/away from Earth. Starting from the layer closest to Earth, they include the Troposphere, Stratosphere, Mesosphere, Ionosphere, and Exosphere.

Troposphere: the lowest layer of Earth's atmosphere. The weather and clouds occur in the troposphere. The troposphere begins at the ground or water level and reaches up to about 11 miles (17 kilometers) high. In the troposphere, the temperature generally decreases as the altitude increases.

Stratosphere: the atmospheric layer between the troposphere and the mesosphere. The stratosphere is characterized by a slight temperature increase with altitude. Only the highest clouds are found in the lowest part of the stratosphere. The temperature slightly increases as altitude increases and clouds decrease. The stratosphere ranges about 11 and 31 miles (17 to 50 kilometers) above Earth's surface. Earth's ozone layer is located in the stratosphere. Ozone is a form of oxygen that is important to survival on Earth because it absorbs the sun's ultraviolet energy.

Mesosphere: the atmospheric layer between the stratosphere and the ionosphere. The mesosphere is characterized by temperatures that quickly decrease as height increases. The mesosphere extends from between 31 and 50 miles (50 to 80 kilometers) above Earth's surface.

Ionosphere: the atmospheric layer between the mesosphere and the exosphere; it is part of the thermosphere. The ionosphere starts at about 43–50 miles (70 to 80 kilometers) high and continues for about 400 miles (640 kilometers). It contains many ions and free electrons. The ions (atoms that have positive or negative changes) are created when sunlight hits atoms and tears off some electrons. Auroras occur in the ionosphere.

Exosphere: the outermost layer of Earth's atmosphere. The exosphere goes from about 400 miles (640 kilometers) high to about 800 miles (1,280 kilometers). The lower boundary of the exosphere is called the critical level of escape, where atmospheric pressure is very low (the gas atoms are very widely spaced) and the temperature is very low.

Step 2

Complete the Checklist "Clues for Success."
The checklist will help you to read and think like a scientist.

Clues for Success

☐ **C**arefully read the information.

☐ **L**ook at any illustrations or diagrams.
They may provide you with additional information to answer the question.

☐ **U**nderstand the way you are asked to answer the question.
- ☐ Graph
- ☐ Chart
- ☐ Diagram
- ☐ Complete sentences
- ☐ Phrase
- ☐ Filled circle

☐ **E**xamine the information given.
- ☐ Reread the questions.
- ☐ Underline key words or phrases.
- ☐ Think about what the questions are asking.

☐ **S**ee if your answers match the questions.

Step 3

Use the information from "Earth's Atmosphere" to complete the graphic organizer.

On the table provided list the layers of the atmosphere. Begin with the layer closest to Earth and work toward the layer closest to outer space. Write the altitude of the layer and a fact about each in the spaces provided.

Atmospheric Layer	Altitude from Earth	Fact

Step 4

Answer the following questions for "Earth's Atmosphere" using information from your graphic organizer.

1. Name the **two** gases that compose most of the atmosphere.

   1. _____

   2. _____

2. List the **two** ways the atmosphere supports the temperature needed for life on Earth.

   1. _____

   2. _____

3. Using complete sentences, explain why each level in the atmosphere is called a sphere.

   _____

   _____

   _____

   _____

4. Why does the article give the measurements for the levels in miles and kilometers?

_____

_____

_____

_____

5. Create a Venn Diagram to compare and contrast the troposphere and stratosphere. Be sure to label the parts of the Venn Diagram.

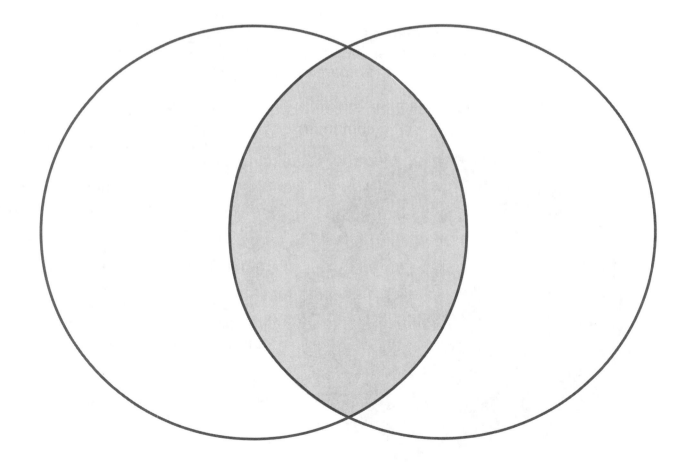

Read the scenario "Earth: More Than Just Soil!"

Earth: More Than Just Soil!

The fifth graders learned that Earth was more than the ground they walked on. In addition to that solid top layer, Earth has several more layers including a center core, a mantel, and the upper crust with plates that are in motion.

Earth's inner core is hot and dense and is composed of iron and nickel. This inner core may be hotter than the sun. Surrounding Earth's inner core is a liquid outer core.

The level surrounding the core is the mantle. The mantle is rocky and composed of the elements silicon, oxygen, magnesium, iron, aluminum, and calcium. The upper part of the mantle is rigid but the inner portion flows slowly moving a few centimeters a year.

The upper layer of Earth is mostly water and a thin, rocky crust which is composed of silicon, aluminum, calcium, sodium, and potassium. This crust is divided into continental plates which drift a few centimeters each year. These plates move over the mantle. They can rub together or pass each other, move apart, over top or under each other. These movements are the beginning of earthquakes, volcanic eruptions, and mountain building. There is a thin crust under the oceans about 6–11 kilometers thick; this is where new crust is formed. The continental crust is about 25–90 kilometers thick.

Step 2

Complete the Checklist "Clues for Success."

The checklist will help you to read and think like a scientist.

Clues for Success

- ☐ **C**arefully read the information.

- ☐ **L**ook at any illustrations or diagrams.
  They may provide you with additional information to answer the question.

- ☐ **U**nderstand the way you are asked to answer the question.
  - ☐ Graph
  - ☐ Chart
  - ☐ Diagram
  - ☐ Complete sentences
  - ☐ Phrase
  - ☐ Filled circle

- ☐ **E**xamine the information given.
  - ☐ Reread the questions.
  - ☐ Underline key words or phrases.
  - ☐ Think about what the questions are asking.

- ☐ **S**ee if your answers match the questions.

Activity 3

Step 3

Use the information from "Earth: More Than Just Soil!" to complete the graphic organizer.

After the class read the description of the layers of Earth, Riki said they reminded her of a hard-boiled egg. She drew this diagram on the board to prove her point.

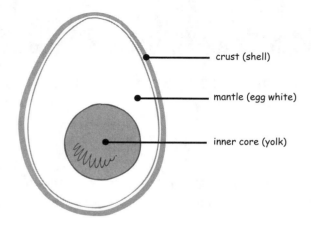

crust (shell)

mantle (egg white)

inner core (yolk)

Compare Riki's model to the information given about the layers of Earth. List three ways they are the same. Then, list three ways Riki's model and the layers of Earth are different.

Same	Different

Activity 3

Step 4

Answer the following questions for "Earth: More Than Just Soil!" using information from your graphic organizer.

Earth's plates can rub together or pass each other, move apart, or move over top or under each other.

1. How do you think these plate movements create earthquakes? Draw **two** Earth plates and use arrows to show the movement of the plates. In the space provided, use a sentence to describe the plate movement in your drawing.

2. How do you think the movements of Earth's plates create mountains? Draw **two** Earth plates and arrows to show movements of the plates. In the space provided, use a sentence to describe the plate movement in your drawing.

Step 1

Read the scenario
"Comets: Dirty Snowballs in Space!"

Comets: Dirty Snowballs in Space!

The fifth grade class was learning about outer space. The students were given the following computer lab research assignment. This assignment will be presented orally to the class.

- Select an outer space topic you are interested in to research.

- Research the topic on science or space related Websites.

- Download an image of your topic. It should be 8" x 10", so the class can see it during your oral presentation.

- Read the information on the Websites and prepare a typed paper containing at least six facts about your topic. Be sure to include your name and resources on your paper.

- Prepare three questions to ask the class after your presentation.

Efraim turned in this work as his assignment.

Comets: Dirty Snowballs in Space!
by Efraim A.

Comets can be called dirty snowballs. They have centers that are mostly solid. This center is called a nucleus. This nucleus is mostly ice and gases with a little dust. The nucleus is surrounded by a thick cloud, called a coma, of water and carbon dioxide. There is also a hydrogen cloud that is very big but it is also very thin. The comet also has a tail that is 10 million kilometers long. It is made up of dust particles. This is the part we see when a comet is visible in the sky. Comets also have an ion tail that can be hundreds of millions of kilometers long.

Comets can only be seen when they are near the sun. The most famous comet is Halley's Comet. The Chinese first recorded seeing the comet in 240 BCE. Halley's comet was last seen in the summer of 1986.

Resources:
www.nasa.gov
www.nineplanets.org

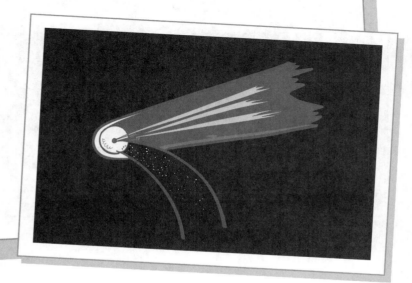

Step 2

Complete the Checklist "Clues for Success."
The checklist will help you to read and think like a scientist.

Clues for Success

☐ **C**arefully read the information.

☐ **L**ook at any illustrations or diagrams.
　　They may provide you with additional information to answer the question.

☐ **U**nderstand the way you are asked to answer the question.
　　☐ Graph
　　☐ Chart
　　☐ Diagram
　　☐ Complete sentences
　　☐ Phrase
　　☐ Filled circle

☐ **E**xamine the information given.
　　☐ Reread the questions.
　　☐ Underline key words or phrases.
　　☐ Think about what the questions are asking.

☐ **S**ee if your answers match the questions.

Step 3

Use the information from "Comets: Dirty Snowballs in Space!" to complete the graphic organizer.

Using the descriptions given in Efraim's report, label the parts of a comet.

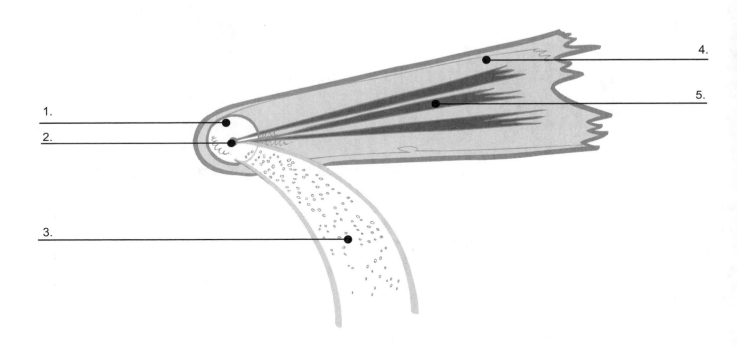

1.

2.

3.

4.

5.

          © Englefield & Associates, Inc.

Step 4

Answer the following questions for "Comets: Dirty Snowballs in Space!" using information from the scenario.

1. Reread Efraim's report. Number and list the facts (true statements) included in his comet report.

Did Efraim include at least six facts about comets in his report?

○ Yes   ○ No

2. Use a complete sentence to answer the questions Efraim asked the class.

Are comets always visible?

_____

_____

_____

Why do you think comets are called dirty snowballs?

_____

_____

_____

Comets are a new discovery in science.

○ True   ○ False

Explain your answer.

_____

_____

_____

_____

3. Review the requirements for the assignment. Mark the ones that Efraim completed in his report.

☐ Select an outer space topic you are interested in to research.

☐ Research the topic on science or space related Websites.

☐ Download an image of your topic. It should be 8" x 10", so the class can see it during your oral presentation.

☐ Read the information on the Websites and prepare a typed paper containing at least six facts about your topic. Be sure to include your name and resources on your paper.

☐ Prepare three questions to ask the class after your presentation.

Did Efraim complete all the requirements for the assignment?

○ Yes   ○ No

Step 1

Read the scenario
"Not All Bad."

Not All Bad

The United States Geological Society depicts natural events affecting the country as not only negative events but also events that may create positive influences on the affected area.

The wildfires that often ravage New Mexico, Colorado, and Southern California create emotional and social problems because residents of the area lost homes, personal belongings, and communities. The natural sites were cleared of dead branches, leaves, needles, and debris. Wildfires replenish the ecosystems with nutrients, such as phosphorous and potassium that are often locked into organisms and ecosystem cycles for years.

Volcanoes are often shown spewing rivers of hot lava and sending out debris that pollute the atmosphere and habitats, and harm organisms including people. Volcanic activity also creates the beginnings and continued growth of new land masses, such as the Galapagos and the Hawaiian Islands. In the Mount Helen area in the northwestern United States, volcanic ash acted as a fertilizer for the regeneration of the forests consumed by volcanic action.

Aquatic areas are also affected by natural events. Two of these natural events that are destructive include hurricanes and landslides. Severe weather conditions, like hurricanes, can devastate cities like New Orleans and cause the collapse of dam and levee systems on the waterways. Landslides created by gravity or weather conditions pile heavy loads of sediment, rocks, and debris into the water. Both of these events also create benefits such as importing fertile soil, nutrients, and minerals to nourish the vegetation and aquatic life. The debris that may harm areas may also create conditions forming new habitats for fish and wildlife.

Step 2

Complete the Checklist "Clues for Success."
The checklist will help you to read and think like a scientist.

Clues for Success

☐ **C**arefully read the information.

☐ **L**ook at any illustrations or diagrams.
They may provide you with additional information to answer the question.

☐ **U**nderstand the way you are asked to answer the question.
☐ Graph
☐ Chart
☐ Diagram
☐ Complete sentences
☐ Phrase
☐ Filled circle

☐ **E**xamine the information given.
☐ Reread the questions.
☐ Underline key words or phrases.
☐ Think about what the questions are asking.

☐ **S**ee if your answers match the questions.

Step 3

Use the information from "Not All Bad" to complete the graphic organizer.

Complete the table below with information about natural events and their impact on surrounding areas.

	Negative Impacts	Positive Impacts
Wildfires		
Volcanoes		
Hurricanes		
Land slides		

 © Englefield & Associates, Inc.

Activity 5

Step 4

Answer the following questions for "Not All Bad" using information from your graphic organizer.

1. Use a complete sentence to tell one way each event described in the scenario helps restore the needs of living organisms.

Wildfires _____

_____

Volcanoes _____

_____

Hurricanes _____

_____

Landslides _____

_____

2. One way a landslide might create new habitats in a river or stream is by depositing boulders or trees. Use a complete sentence to give an example of how depositing boulders or trees could help create new habitats.

_____

_____

_____

_____

_____

Chapter 5

The activities in this section of the book will focus on Science and Technology.

These activities will help you develop:

- abilities of technological design and
- understandings about science and technology.

Use the "Clues for Success" Checklists as you complete each activity in this section as a tool to help you do your best work.

Step 1

Read the scenario "Science Helps."

Science Helps

Children and adults have questions about the world. Science is a way of looking at the world and finding answers to questions or explaining the natural world. Sometimes the information needed can be found in books, on the Internet, or asking a scientist. Other times, you can conduct and experiment.

Information such as the temperature water boils in degrees Celsius can be found in books or on the Internet.

The change in the area of the rainforest or current weather conditions can be found on the Internet.

Often information about the environment or a medical condition can be obtained by asking a scientist.

Sometimes, students and adults conduct experiments to find out answers or create solutions to problems.

Step 2

Complete the Checklist "Clues for Success."

The checklist will help you to read and think like a scientist.

Clues for Success

☐ **C**arefully read the information.

☐ **L**ook at any illustrations or diagrams.
They may provide you with additional information to answer the question.

☐ **U**nderstand the way you are asked to answer the question.
- ☐ Graph
- ☐ Chart
- ☐ Diagram
- ☐ Complete sentences
- ☐ Phrase
- ☐ Filled circle

☐ **E**xamine the information given.
- ☐ Reread the questions.
- ☐ Underline key words or phrases.
- ☐ Think about what the questions are asking.

☐ **S**ee if your answers match the questions.

Step 3

Use the information from "Science Helps" to complete the graphic organizer.

Lee is involved in several sports in school. The result is a messy room that has piles of volleyball and softball uniforms, shoes, team schedules, and sports equipment.

Outline a plan to help with the organization of Lee's room.

I. Uniform organization

    A. _____

    B. _____

I. Team schedule organization

    A. _____

    B. _____

I. Sports equipment organization

    A. _____

    B. _____

Activity 1

Step 4

Answer the following questions for "Science Helps" using information from your graphic organizer.

1. Select **two** items from the list. Then use a sentence to describe how you would use them to solve the problem of Lee's messy room.

○ shoe boxes   ○ bulletin board

○ plastic tubs   ○ hangers

○ baskets   ○ trash can

item:_____

use:_____

_____

item:_____

use:_____

_____

Step 1

Read the scenario "Studying a Scientist."

Studying a Scientist

The students watched a DVD about George Washington Carver. Each segment of the DVD had a sub-heading about a specific time in Carver's life. After each portion, the students were challenged to write a summary about the material presented. Here is Michalle's summary for each segment.

George Washington Carver (GWC)

The Early Days
GWC was born to slaves in Missouri. He had whopping cough and was a sick child. His chores were cooking, sewing, and working in the garden. He started school at the age of 12. He taught himself to read. He didn't have a pencil, so he made himself a tool to hold pieces of pencils to be able to write.

The Academic Years
GWC was turned away from college because he was black. At thirty, he was finally accepted to college. He studied botany (study of plants). He didn't have much money. He earned money by cooking for friends and ironing clothes for his friends. He made himself notebooks from old wrapping paper. After he graduated, he taught biology to beginning students. Booker T. Washington convinced Carver to teach at the Tuskegee Normal and Industrial Institute for Negroes in Alabama where he headed the agricultural department for 50 years. He used glass bottles from the dump for experiments because they did not have scientific equipment.

Contributions

GWC wanted to help poor farmers. He introduced the peanut as an alternative to the cotton plant because replanting cotton was using all of the nutrients from the soil. He also studied diseases that harmed the cotton crop. He developed over 300 uses for the peanut to show farmers that it was a crop that would have many uses including soups, soap, and shaving cream. He also taught people many new uses for sweet potatoes, pecans, and soy beans.

Summary

GWC was a humble man. He donated his life savings to the Tuskegee University. He was not interested in wearing stylish clothes. He usually wore a wrinkled suit with a fresh flower pinned on it. He lived his life to improve the life of farmers and serve others.

He did not invent peanut butter!

Activity 2

Step 2

Complete the Checklist "Clues for Success."

The checklist will help you to read and think like a scientist.

Clues for Success

☐ **C**arefully read the information.

☐ **L**ook at any illustrations or diagrams.
They may provide you with additional information to answer the question.

☐ **U**nderstand the way you are asked to answer the question.
☐ Graph
☐ Chart
☐ Diagram
☐ Complete sentences
☐ Phrase
☐ Filled circle

☐ **E**xamine the information given.
☐ Reread the questions.
☐ Underline key words or phrases.
☐ Think about what the questions are asking.

☐ **S**ee if your answers match the questions.

Step 3

Use the information from "Studying a Scientist" to complete the graphic organizer.

The teacher told the students that George Washington Carver demonstrates many qualities that are important for scientists. She listed several on a chart and challenged the students to show how Carver showed those traits. Use the information from Michalle's summary to help you complete the chart.

Quality of a Scientist	Carver's Life
Reasoning	
Insight	
Energy	
Skill	
Creativity	
Openess to New Ideas	

Step 4

Answer the following questions for "Studying a Scientist" using information from your graphic organizer.

Three qualities of scientists are listed. Use an example to illustrate why that quality is important for a scientist.

1. Skill _____

_____

_____

2. Creativity _____

_____

_____

3. Openness to new ideas _____

_____

_____

Step 1

Read the scenario "Reporting on a Scientist."

Reporting on a Scientist

The fifth grade class was doing biographical reports on scientist. Jacob was interested in computers and decided to research scientists that were working in the field of computer technology. He decided that he would research the World Wide Web (WWW).

In his search, Jacob learned the primary inventor of the WWW was Tim Berners-Lee. Here is the information Jacob presented to the class.

Tim Berners-Lee: Inventor of the World Wide Web

Tim Berners-Lee was born on June 8, 1955, in London, England. His parents taught him that learning mathematics was very important. They even quizzed Tim with math problems when the family was out having fun and eating dinner. His parents worked together to help develop one of the earliest computers.

In 1976, Berners-Lee graduated from The Queen's College in Oxford, England, with a degree in physics. While he was there, he got in trouble for breaking into (hacking) the university's computer system. Berners-Lee also built a computer from electronic equipment he found including an old television. After graduation, he worked as a computer programmer and wrote software for computers.

At the time, Berners-Lee was working at CERN, the European Particle Physics Laboratory in Geneva, Switzerland. He invented a system of sharing scientific data (and other information) around the world, using the Internet, a world-wide network of computers and hypertext documents. Berners-Lee developed the work for some of the common computer terms used today. He wrote the language HTML (HyperText Mark-up Language), URLs (universal resource locators) to designate the location of each Web page, and HTTP (HyperText Transfer Protocol) the rules for linking to pages on the Web.

Here is how Berners-Lee developed and worked on this project. 1980—Berners-Lee proposed a project based on the concept of hypertext, to facilitate sharing and updating information among researchers. This was his first version of the Web, a program named "Enquire Within Upon Everything." It was later shortened to "Enquire."

1989—At CERN, the largest Internet facility in Europe, Berners-Lee saw an opportunity to join hypertext with the Internet creating the World Wide Web. He began working on his first proposal in March 1989.

                   © Englefield & Associates, Inc.

1990—With the help of Robert Cailliau, Berners-Lee produced a version of the World Wide Web that was acceptable to the scientific community using similar ideas to his "Enquire" system. Berners-Lee designed and built the first Web browser, editor, and the first Web server.

1991—At CERN, the first Web site was built and put online. This first site explained what the World Wide Web was, how to own a browser, and how to set up a Web server.

1994—Berners-Lee founded the World Wide Web Consortium (W3C) at the Massachusetts Institute of Technology. It is made up of companies that are willing to support and work to improve the quality of the Web.

It is important to know that Berners-Lee obtained no patent (legal rights) for his work and receives no royalties (money) for his ideas. He and W3C want the ideas to be available and used by the public. In 2003, Queen Elizabeth II knighted Berners-Lee for his work on the Web.

Step 2

Complete the Checklist "Clues for Success."

The checklist will help you to read and think like a scientist.

Clues for Success

☐ **C**arefully read the information.

☐ **L**ook at any illustrations or diagrams.
They may provide you with additional information to answer the question.

☐ **U**nderstand the way you are asked to answer the question.
　☐ Graph
　☐ Chart
　☐ Diagram
　☐ Complete sentences
　☐ Phrase
　☐ Filled circle

☐ **E**xamine the information given.
　☐ Reread the questions.
　☐ Underline key words or phrases.
　☐ Think about what the questions are asking.

☐ **S**ee if your answers match the questions.

Step 3

Use the information from "Reporting on a Scientist" to complete the graphic organizer.

Complete the outline for the report Jacob presented to the class.

Tim Berners-Lee

I. Childhood Experience

Details _____

_____

II. College Experience

Details _____

_____

III. Systems Invented Used Today

HTML _____

URL _____

HTTP _____

IV. Timeline

1989 _____

1990 _____

1991 _____

1994 _____

2003 _____

Step 4

Answer the following questions for "Reporting on a Scientist" using information from your graphic organizer.

1. List **three** statements about the qualities of an inventor/scientist learned from the life of Tim Berners-Lee.

    1. _____

    _____

    2. _____

    _____

    3. _____

    _____

2. Tim Berners-Lee did not work/invent alone. List **two** pieces of information from the report showing that Berners-Lee worked with others.

    1. _____

    _____

    2. _____

    _____

3. Tim Berners-Lee thought it was important for everyone to have free access to his work, so he did not patent or accept royalties for his work. Why do you think this was important to him?

_____

_____

_____

_____

_____

_____

_____

_____

Activity 4

Step 1

Read the scenario
"Birds, Bicycles, and Biplanes."

Birds, Bicycles, and Biplanes

Wilbur and Orville Wright were brothers that made a remarkable contribution to society. They were the inventors that designed and developed the first flying machine. Although they did not have any formal academic training, they used their talents and observations to become successful scientists.

As young boys growing up in Ohio, the brothers followed the inquisitive nature of their mother. She enjoyed exploring mechanical instruments and toys. The boys enjoyed playing with mechanical objects as well. In addition, they had many jobs and hobbies that contributed to their background in engineering, including building and repairing bicycles and constructing a printing press. They were self taught business men that used their profits and talents to fund their flying adventures.

In addition, the brothers observed the aerodynamic properties of birds and incorporated these ideas into their inventions.

In 1900, they designed and built a glider that cost $15. Their first invention was not a success and did not glide or fly. They constructed a wind tunnel to help them rework and improve their design. They built over 200 models. In 1902, a glider was successfully airborne at Kill Devil's Hill near Kitty Hawk, North Carolina.

After several unsuccessful attempts followed by adjustments and repairs, the Wright brothers successfully tested a two propeller plane with a 12 horse, internal combustion engine in December of 1903. In its first flight, the plane flew 120 feet and was in the air for 12 seconds.

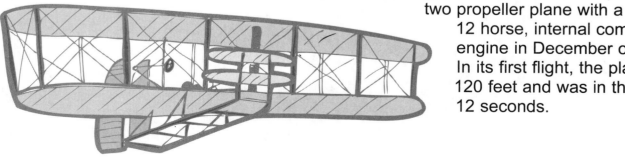

Step 2

Complete the Checklist "Clues for Success."

The checklist will help you to read and think like a scientist.

Clues for Success

☐ **C**arefully read the information.

☐ **L**ook at any illustrations or diagrams.
They may provide you with additional information to answer the question.

☐ **U**nderstand the way you are asked to answer the question.
☐ Graph
☐ Chart
☐ Diagram
☐ Complete sentences
☐ Phrase
☐ Filled circle

☐ **E**xamine the information given.
☐ Reread the questions.
☐ Underline key words or phrases.
☐ Think about what the questions are asking.

☐ **S**ee if your answers match the questions.

Activity 4

Step 3

Use the information from "Birds, Bicycles, and Biplanes" to complete the graphic organizer.

The Wright Brothers were inspired to invent their flying machine from a variety of areas. Several of these areas are listed. Consider each cause and explain the effect that it contributed to the Wright brothers' invention.

Cause	Effect
Birds in flight	
Constructing mechanical toys	
Building bicycles	
Failed flights of inventions	

Step 4

Answer the following questions for "Birds, Bicycles, and Biplanes" using information from your graphic organizer.

Consider the following inventions. Name and explain how an earlier invention made development of each modern technology possible.

1. Skate Board

Invention _____

Explanation _____

_____

_____

_____

2. Cell phone

Invention _____

Explanation _____

_____

_____

_____

_____

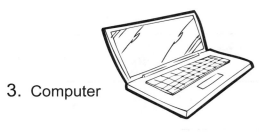

3. Computer

Invention _____

Explanation _____

_____

_____

_____

_____

Chapter 6

Science in Personal and Social Perspective

The activities in this section of the book will focus on Science in Personal and Social Perspective.

The activities in this section will help you focus on:

- personal health,

- populations, resources, and environments,

- natural hazards,

- risks and benefits, and

- science and technology in society.

Use the "Clues for Success" Checklists as you complete each activity in this section as a tool to help you do your best work.

Step 1

Read the scenario "Spinning Senses."

Spinning Senses

Life Science

The behavior of individual organisms is influenced by internal cues (such as hunger) and external cues (such as changes in the environments). Humans have senses that help them detect internal and external cues.

These interactions might include:

Smelling the odor of a pine tree.

Tasting sweet orange juice.

 Hearing the sounds of nature.

 Seeing the colors of a rainbow.

 Touching the soft fur of a guinea pig.

 © Englefield & Associates, Inc.

Step
2

Complete the Checklist "Clues for Success."

The checklist will help you to read and think like a scientist.

Clues for Success

☐ **C** arefully read the information.

☐ **L** ook at any illustrations or diagrams.
   They may provide you with additional information to answer the question.

☐ **U** nderstand the way you are asked to answer the question.
   ☐ Graph
   ☐ Chart
   ☐ Diagram
   ☐ Complete sentences
   ☐ Phrase
   ☐ Filled circle

☐ **E** xamine the information given.
   ☐ Reread the questions.
   ☐ Underline key words or phrases.
   ☐ Think about what the questions are asking.

☐ **S** ee if your answers match the questions.

Step 3

Use the information from "Spinning Senses" to complete the graphic organizer.

The class played a game with the spinner shown. The children's responses to the spinner game are listed below. Decide where the spinner pointed for each of the answers listed. Write the correct sense in the space provided. The first one has been done for you.

Childrens' Answers	Spinner Position
The scent of a flower	**smell**
Drinking a hot cup of cocoa	
The slippery skin of a snake	
A loud whistle	
A beautiful rainbow	

Step 4 — Answer the following questions for "Spinning Senses" using information from your graphic organizer.

1. Read the descriptions below and determine if the cue described is an **external** cue or an **internal** cue. Put an X in the correct box.

	External Cues from the environment	Internal Cues from your body
Billy was tired after the softball game.		
Chris wanted to eat some of the freshly baked cookies that she smelled.		
Joan was thirsty when she woke up in the morning.		
George heard the lawn mower outside his window.		
Jacob laughed when he watched cartoons.		
Tom was excited when he went to the amusement park.		
Martha felt shy on the first day at her new school.		

Step 1

Read the scenario "A Food Pallet of Colors."

A Food Pallet of Colors

Nurse Cathy came to the fifth grade class to speak about nutrition. She was going to teach a six-week course to help the students learn about good eating habits. Nurse Cathy told them that fruits are good sources of potassium and fiber. Citrus fruits, melons, and berries are especially good sources of vitamin C. And all yellow fruits are rich sources of vitamin A. She brought several fruits and vegetables for the students to try. First they tried to guess fruits by their smell and feel. After the game, the students enjoyed sampling fruits from mango to ugly fruit. While the students were munching on carrots, celery, and cherry tomatoes, the nurse said they should eat dark green, leafy and orange vegetables, and dry beans and peas several times a week. Dark green, leafy vegetables are good sources of vitamins A and C, calcium, magnesium, potassium, and fiber. Orange vegetables are excellent sources of vitamin A. Dry beans and peas are good sources of fiber, potassium, protein, starch, and minerals. Other vegetables contain varying amounts of vitamins, minerals, and fiber.

Nurse Cathy spoke about the importance of variety and serving size for fruits and vegetables.

She told the students about some of the serving sizes included in the chart below.

6 ounces of fruit or vegetable juice	
1/2 cup of dried fruit	1 pear
1 cup of salad greens	1 ear of corn
6 slices of canned peaches	6 strawberries
20 grapes	10 string beans

She challenged the students to eat a variety of five fruits and vegetables each day for a week.

Students' choices could be fresh, frozen, dried, or canned fruits and vegetables. They should keep a chart of the fruits and vegetables they ate. All of the students meeting their fruit and vegetable requirements would receive a ticket to a frozen-yogurt party.

Nathaniel wanted to go to the yogurt party, so he planned on recording all the fruits and vegetables he ate. Here is his log for the first two days of the week.

Monday	
Breakfast	waffle orange juice
Lunch	peanut butter and banana sandwich apple juice piece of cantaloupe
Dinner	veggie pizza with peppers, mushrooms, onions (2 pieces) small (1 cup) green salad with Italian dressing
Snack	1 small orange with yogurt dip
Tuesday	
Breakfast	cereal with 1 small banana
Lunch	cheese and crackers with grapes apple juice
Dinner	string beans cole slaw (cabbage salad)
Snack	1 cup of fresh pineapple

Step 2

Complete the Checklist "Clues for Success."
The checklist will help you to read and think like a scientist.

Clues for Success

- [ ] **C**arefully read the information.

- [ ] **L**ook at any illustrations or diagrams.
  They may provide you with additional information to answer the question.

- [ ] **U**nderstand the way you are asked to answer the question.
  - [ ] Graph
  - [ ] Chart
  - [ ] Diagram
  - [ ] Complete sentences
  - [ ] Phrase
  - [ ] Filled circle

- [ ] **E**xamine the information given.
  - [ ] Reread the questions.
  - [ ] Underline key words or phrases.
  - [ ] Think about what the questions are asking.

- [ ] **S**ee if your answers match the questions.

Step 3

Use the information from "A Food Pallet of Colors" to complete the graphic organizer.

Complete the chart to show the number and kinds of fruits and vegetables Nathaniel ate Monday and Tuesday.

	Monday	Tuesday
Fruits		
Vegetables		
TOTAL		

Did Nathaniel meet the five fruit and vegetable requirement for Monday?

○ yes   ○ no

Did Nathaniel meet the five fruit and vegetable requirement for Tuesday?

○ yes   ○ no

Answer the following questions for "A Food Pallet of Colors" using information from your graphic organizer.

Nurse Cathy also mentioned that the students should eat a variety of different "colors" of fruits and vegetables.

1. Look at the list of fruit and vegetable colors Nurse Cathy mentioned. Then reread Nathaniel's food log for Monday and Tuesday to show which "colors" of these foods Nathaniel ate.

	Monday	Tuesday
Red		
Yellow/Orange		
Green		
White		
Blue/Purple		

 © Englefield & Associates, Inc.

2. Did Nathaniel eat fruits and vegetables from each color group?

   O yes   O no

3. What fruits or vegetables could Nathaniel add to his diet to include all color groups?

   fruits:_____

   vegetables:_____

4. During the class the nurse mentioned that eating an orange was better than drinking orange juice. What do you think she meant when she said that?

   _____

   _____

5. Do you eat a variety of fruits and vegetables?

   O yes   O no

6. What is your favorite fruit or vegetable?

   _____

   Do you like to eat it cooked or raw?

   O cooked   O raw

Step 1

Read the scenario "Groundwater Awareness."

Groundwater Awareness

When they were asked about pollution the students quickly identified air pollution. At the Environmental Day assembly, the speaker asked the students to help their community learn more about groundwater contamination. Here is some of the information she shared with them.

Groundwater is the part of the water cycle that is often forgotten because it is held within the pores and cracks between the particles of the earth like dirt and sand. Groundwater is important because fifty percent of the people in United States depend on groundwater for daily drinking water. Farmers also use groundwater as an important source of irrigation water.

Groundwater contamination occurs when man-made products, such as gasoline, oil, road salts, and chemicals, seep into the water held under ground and make it unsafe and unfit for use by humans and other parts of nature. Some contaminants include storage tanks, septic systems, hazardous waste sites, and landfills. In addition, pollutants that contaminate surface water like creeks, rivers, and lakes move through the soil and end up in the groundwater. Some examples include pesticides and fertilizers, road salt, and harmful substances from home, industrial, and mining sites.

At the end of her presentation, she challenged the students to help create awareness about ground water contamination.

Step 2

Complete the Checklist "Clues for Success."

The checklist will help you to read and think like a scientist.

Clues for Success

☐ **C** arefully read the information.

☐ **L** ook at any illustrations or diagrams.
    They may provide you with additional information to answer the question.

☐ **U** nderstand the way you are asked to answer the question.
    ☐ Graph
    ☐ Chart
    ☐ Diagram
    ☐ Complete sentences
    ☐ Phrase
    ☐ Filled circle

☐ **E** xamine the information given.
    ☐ Reread the questions.
    ☐ Underline key words or phrases.
    ☐ Think about what the questions are asking.

☐ **S** ee if your answers match the questions.

Step 3

Use the information from
"Groundwater Awareness"
to complete the graphic organizer.

Troy was selected to represent the class and talk to the mayor about groundwater. Create a summary of the information for him to use.

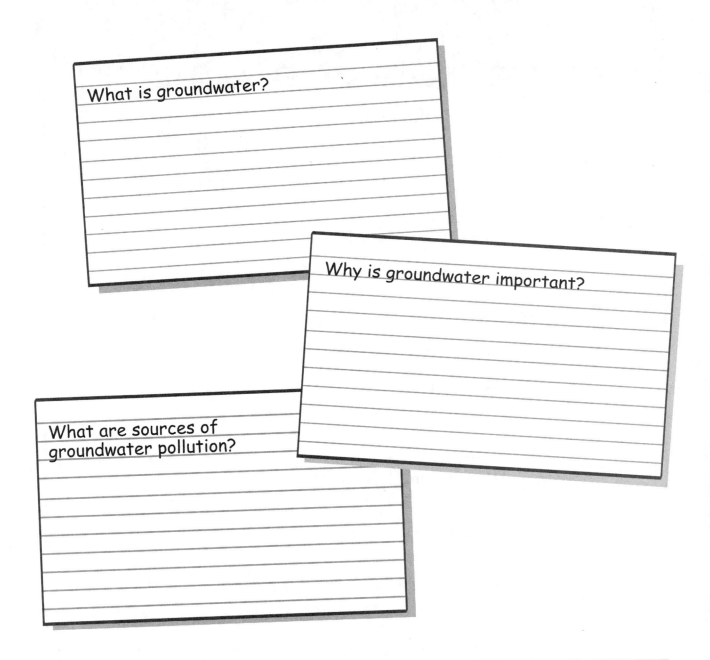

What is groundwater?

Why is groundwater important?

What are sources of groundwater pollution?

         © Englefield & Associates, Inc.

Step 4

Answer the following questions for "Groundwater Awareness" using information from your graphic organizer.

1. List **three** ways the students might create awareness about groundwater.

    1. _____

    2. _____

    3. _____

2. Groundwater is a part of the water cycle. Illustrate and label the water cycle. Use arrows to show the movement of the water through the processes of precipitation, condensation, and evaporation. Put a star by the area of groundwater.

Activity 4

Step 1

Read the scenario "To Skateboard or Not."

To Skateboard or Not

George loves to skateboard! At his parents' request, the school nurse considered the possibility of George skateboarding to school. At first Nurse Tom discussed several topics with the family. First, they discussed the potential hazards present on the route George would follow each day. Second, they discussed the need for injury prevention with the use of personal safety gear. Third, George must promise to be careful, if permitted to ride his skateboard to school.

With the help of the community sports center professional, George, his family, and the nurse had a safety consultation. At the center, George heard about various safety equipment and why it is important.

Here is a poster George brought back to share with the class.

Activity 4

Step 2

Complete the Checklist "Clues for Success."
The checklist will help you to read and think like a scientist.

Clues for Success

☐ **C**arefully read the information.

☐ **L**ook at any illustrations or diagrams.
They may provide you with additional information to answer the question.

☐ **U**nderstand the way you are asked to answer the question.
   ☐ Graph
   ☐ Chart
   ☐ Diagram
   ☐ Complete sentences
   ☐ Phrase
   ☐ Filled circle

☐ **E**xamine the information given.
   ☐ Reread the questions.
   ☐ Underline key words or phrases.
   ☐ Think about what the questions are asking.

☐ **S**ee if your answers match the questions.

Step 3

Use the information from "To Skateboard or Not" to complete the graphic organizer.

Consider each injury listed below. Then decide which piece of safety equipment might help prevent the injury.

To Prevent This ➡	Use this equipment
Foot injury ➡	
Head injury ➡	
Concussion ➡	
Inflammation of the knees ➡	
Wrist fractures ➡	
Permanent brain damage ➡	
Joint dislocation ➡	
Sprained ankle ➡	
Broken bones ➡	

 © Englefield & Associates, Inc.

Step 4

Answer the following questions for "To Skateboard or Not" using information from your graphic organizer.

1. The poster suggested wearing reflective or bright clothing. Why is this important for a skateboarder?

    _____

    _____

    _____

2. The nurse and the sport's specialist were focusing on George's personal safety precautions. Use complete sentences to list **two** safety issues that George will not be able to control.

    1. _____

       _____

    2. _____

       _____

3. The final decision was to deny George's request to ride his skateboard to school. Do you agree or disagree with this decision?

○ agree   ○ disagree

Support your opinion with **three** ideas based on the information in the scenario or your knowledge about skateboarding.

1. _____

_____

_____

2. _____

_____

_____

3. _____

_____

_____

Step 1

Read the scenario "Indoor Air Pollution."

Indoor Air Pollution

Andrew and Jonah were researching indoor air pollution for a health class presentation. They were very interested in the topic since they were studying sources of air pollution in science class. Their presentation was divided into two main sections.

The Concern about Indoor Air Pollution

Since the early 1970s buildings have been built to be more airtight to conserve energy. This has resulted from using improved construction techniques and caulking and sealing. Unfortunately, these construction improvements limit the amount of polluted air that escapes. This situation can cause pollutants to build up to unhealthy levels inside a house, school, or other building. In addition to allergens being trapped in homes by better seals on windows and doors, the chemicals used in our homes for everyday cleaning and maintenance also cause indoor air pollution problems. These chemicals can build up over time, eventually reaching harmful levels when the air in a home is not moving and is not purified. Dust, pollen from plants, and pet dander can also contaminate the air.

Health Risks Caused by Indoor Air Pollution

The Environmental Protection Agency stated that poor indoor air quality is one of the top risks to public health. Indoor air pollution may cause short or long-term health effects. The short-term problems could include symptoms such as burning eyes, skin irritation, and headaches. Symptoms might progress to sneezing, coughing, shortness of breath, fever, and dizziness. In addition, other infections such as influenza, measles, and chicken pox are also transmitted through the air.

Asthma is a specific condition affected by indoor pollution. Asthma is an inflammation (redness, irritation, swelling) of the bronchial airways. It can be caused by many factors including smoke, pet dander, pollen, and other tiny particles found inside the home. Asthma is also produced by allergies, respiratory infections, and airborne irritants. An asthma attack occurs when the normal function of the air passages over-react producing more mucus, swelling, and more muscle contractions. The signs of asthma are chest tightness, coughing, and wheezing. These can lead to shortness of breath and low blood oxygen.

Activity 5

Step 2

Complete the Checklist "Clues for Success."
The checklist will help you to read and think like a scientist.

Clues for Success

☐ **C**arefully read the information.

☐ **L**ook at any illustrations or diagrams.
They may provide you with additional information to answer the question.

☐ **U**nderstand the way you are asked to answer the question.
   ☐ Graph
   ☐ Chart
   ☐ Diagram
   ☐ Complete sentences
   ☐ Phrase
   ☐ Filled circle

☐ **E**xamine the information given.
   ☐ Reread the questions.
   ☐ Underline key words or phrases.
   ☐ Think about what the questions are asking.

☐ **S**ee if your answers match the questions.

Activity 5

Step 3

Use the information from "Indoor Air Pollution" to complete the graphic organizer.

List **three** things that may cause Indoor air pollution.

These conditions → Indoor Air Pollution	
	Indoor Air Pollution
	Indoor Air Pollution
	Indoor Air Pollution

Indoor air pollution may cause a variety of health conditions. List **three** health conditions in the spaces provided.

Indoor Air Pollution → These health conditions	
Indoor Air Pollution	
Indoor Air Pollution	
Indoor Air Pollution	

Step 4

Answer the following questions for "Indoor Air Pollution" using information from your graphic organizer.

1. Indoor air pollution may cause symptoms such as burning eyes, skin irritation, headaches, coughing, sneezing, shortness of breath, and wheezing. These effects impact the systems of the body.

   Name **two** systems of the body that are impacted by indoor air pollution. Then, explain why.

   System _____

   Reason _____

   _____

   _____

   System _____

   Reason _____

   _____

   _____

2. List **two** ways you can make the air in your home cleaner.

   1. _____

   2. _____

3. Asthma is a long term condition that may be caused by indoor air pollution. Describe how asthma impacts the body.

   _____

   _____

   _____

   _____

   _____

   _____

Chapter 7

The activities in this section of the book will focus on the History and Nature of Science.

The activities in this section will help you learn:

- science as a human endeavor,

- nature of science, and

- history of science.

Use the "Clues for Success" Checklists as you complete each activity in this section as a tool to help you do your best work.

Step 1

Read the scenario
"Inventing Toys."

Inventing Toys

These toys were arranged on the four lab tables for a Friday special.
Each toy had a card with information about the inventor, date of
invention, and one "Fun Fact," a special piece of information about
the toy.

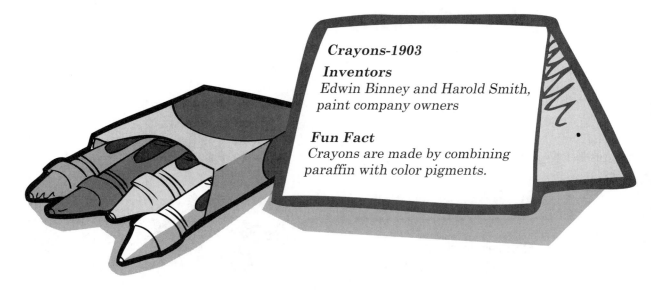

Crayons-1903

Inventors
Edwin Binney and Harold Smith,
paint company owners

Fun Fact
Crayons are made by combining
paraffin with color pigments.

LEGO ™—1958

Inventor
Godtfred Christiansen,
Danish carpenter

Fun Fact
The word LEGO™ comes from the
Danish phrase meaning play well,
"LEg GOdt".

Play-Doh—1956

Inventors
Noah and Joseph McVicker,
owners of a soap company

Fun Fact
Originally developed as a wall
paper cleaner, the only color
was white.

Slinky—1943

Inventor
Richard James, marine engineer

Fun Fact
The slinky was accidently invented
when James, an engineer, was
developing springs to test the
power of equipment on a ship.

K'nex—1990

Inventors
Joel Glickman,
industrial engineer

Fun Fact
Glickman got the idea to invent
K'nex while he was playing with
straws at a wedding.

Step 2

Complete the Checklist "Clues for Success."
The checklist will help you to read and think like a scientist.

Clues for Success

☐ **C**arefully read the information.

☐ **L**ook at any illustrations or diagrams.
They may provide you with additional information to answer the question.

☐ **U**nderstand the way you are asked to answer the question.
☐ Graph
☐ Chart
☐ Diagram
☐ Complete sentences
☐ Phrase
☐ Filled circle

☐ **E**xamine the information given.
☐ Reread the questions.
☐ Underline key words or phrases.
☐ Think about what the questions are asking.

☐ **S**ee if your answers match the questions.

Step 3

Use the information from
"Inventing Toys"
to complete the graphic organizer.

Create a timeline using the information about the toys found in the lab.

Be sure to create a timeline with equal units. Then, list the toy and the year it was invented.

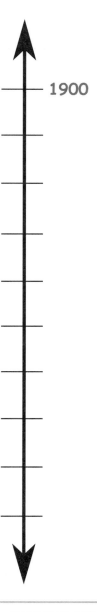

1900

> ## Step 4
>
> ## Answer the following questions for "Inventing Toys" using information from the scenario and your graphic organizer.

Toys are important inventions. Reread the original scenario, then use a complete sentence to answer the questions below.

1. Which of the toys described were discovered while the inventors were trying to develop another product?

_____

_____

_____

_____

2. Two of the toys were developed by business owners. List the toys, then describe how the businesses might have helped the owners create the toys.

Toy 1 _____

Explain how the business might have helped develop the toy.

_____

_____

_____

Toy 2 _____

Explain how the business might have helped develop the toy.

_____

_____

_____

_____

3. List **two** ways you think learning about how science works might have helped these inventors.

   1. _____

_____

_____

   2. _____

_____

_____

Step 1

Read the scenario "Scientific Theory."

Scientific Theory

The students were learning about scientific theory. They thought scientific theory meant a hunch or a guess. Their teacher explained that scientific theories are more than a hunch. Scientific theories are developed by making observations based on experiments or models, then presenting evidence to support why something works in a particular way. Many scientists must agree on the information before it is accepted as a scientific theory. Sometimes these ideas can be changed when a scientist learns new information about a topic.

The teacher wrote these statements about scientific theories on the board for the students.

> Theories include observations.
> Theories have been tested many times.
> Theories are inferred explanations, strongly supported by evidence.
> Theories are used to make predictions.

She then challenged them to read the information about the theory of plate tectonics in their monthly science magazine.

Plate Tectonics

The study of plate tectonics helps answer questions about the history of Earth by looking at the activity of the movement of Earth's crust and mantle. Scientists look for answers to questions like:

- How could the same organisms have lived at the same time on different continents?
- What actions are responsible for the creation of mountains and the actions of volcanoes and earthquakes?

This plate tectonics theory combines ideas about the motion of the crust and mantle of Earth from Alfred Wegener's theory of continental drift and ideas about how the sea floor spreads.

Wegener's theory supported the idea that Earth was a large land mass that slowly drifted apart on top of a liquid layer of the ocean floor. The evidence for continental drift is now extensive. Similar plants, animals, and fossils are found on different continents leading to the idea that continents were once connected. There is also living evidence of the same animals found on two separate continents. Scientists did have a problem with Wegener's theory; he believed that the continents moved across the top layer of the ocean floor.

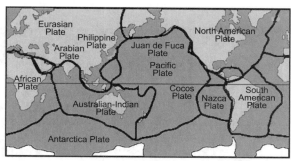

Major Tectonic Plates of the World

The theory of the sea floor spreading supported the idea that new surface of Earth is created by volcanoes at the ocean ridges and returned to Earth's mantle through the ocean trenches. Here, new surface is created and the old surface is recycled back into the mantle.

On Earth's surface, tectonic activity takes place at the plate boundaries where a new surface is formed; an old surface is returned to the mantle, two land masses collide; and two plates rub along each other.

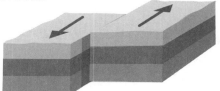

A transform boundary is one way that tectonic plates shift.

Step 2

Complete the Checklist "Clues for Success."

The checklist will help you to read and think like a scientist.

Clues for Success

- ☐ **C**arefully read the information.

- ☐ **L**ook at any illustrations or diagrams.
  They may provide you with additional information to answer the question.

- ☐ **U**nderstand the way you are asked to answer the question.
  - ☐ Graph
  - ☐ Chart
  - ☐ Diagram
  - ☐ Complete sentences
  - ☐ Phrase
  - ☐ Filled circle

- ☐ **E**xamine the information given.
  - ☐ Reread the questions.
  - ☐ Underline key words or phrases.
  - ☐ Think about what the questions are asking.

- ☐ **S**ee if your answers match the questions.

Step 3

Use the information from "Scientific Theory" to complete the graphic organizer.

Using the information about plate tectonics. Complete the following chart.

	Continental drift	Sea floor spreading
Briefly state the theory		
Evidence that supports theory		
Predictions made using the theory		

Activity 2

Step 4

Answer the following questions for "Scientific Theory" using information from your graphic organizer.

1. What would happen if a new ocean floor was made, but the old ocean floor was not returned to the mantle?

_____

_____

_____

_____

2. Scientists combined two theories into one on plate tectonics. List **two** reasons why this was necessary.

   1. _____

     _____

   2. _____

     _____

3. Why do you think scientists did not agree that continents rubbed against the ocean floor when they drifted apart?

_____

_____

_____

_____

_____

_____

Activity 3

Step 1

Read the scenario
"Water on Mars!"

Water on Mars!

The students learned that on July 31, 2008, NASA scientists confirmed that there was water on Mars. A tiny oven on the NASA spacecraft Phoenix Lander heated a scoop of Martian dirt that gave off some traces of water vapor! This came after many years of scientists developing ideas, tools, and techniques to make learning about water on Mars possible. These technologies made it possible for the Phoenix Lander to get the information (data) to confirm the theory.

Technological developments make all the missions to Mars possible. Each mission is built on the information gained from past missions. New technologies are developed to meet the needs of every mission. Different types of technological developments lead to the discovery of water on Mars. Some of these categories included:

- Energy to get to Mars, maintain the systems on the equipment, and conduct the missions,

- Communication systems and software to support commands and transmit data,

- Creating systems to handle the environmental conditions on Mars,

- Equipment and collection tools to examine samples on the planet's surface and to transport samples to Earth for more studies, and

- Cleaning and sterilizing spacecraft.

Mission highlights before July 31, 2008 that lead to the discovery of water on Mars include:

- Information from Mariner 4 (1965) showed that light and dark patches seen on Mars were not water markings as some scientists thought.

- In 2002, observations by an orbiting probe found evidence of vast amounts of water locked up in ice below the surface.

- Mars Express and the Mars Reconnaissance Orbiter revealed data illustrating large quantities of water as ice both at the poles (2005) and at the mid-latitudes of the planet (2008).

- Evidence from the planet's surface gathered by the Phoenix Lander and earlier missions suggests that Mars previously had large-scale water coverage. Some observations lead scientists to believe that small water flows might have spouted up during the past decade.

- Scientists continue to develop new tools and ways of getting evidence about Mars. Two rovers, Spirit and Opportunity, have been created to explore the surface of the planet. Balloons and airplanes are being developed to get closer and better images of the surface of Mars and cover more area faster than today's equipment!

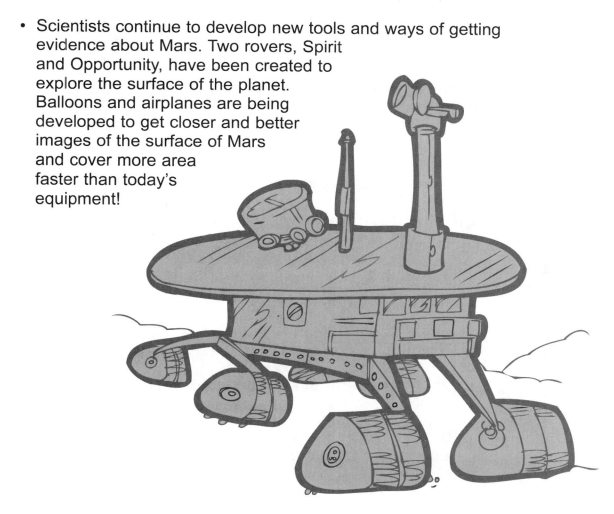

Step 2

Complete the Checklist "Clues for Success."

The checklist will help you to read and think like a scientist.

Clues for Success

☐ **C**arefully read the information.

☐ **L**ook at any illustrations or diagrams.
They may provide you with additional information to answer the question.

☐ **U**nderstand the way you are asked to answer the question.
☐ Graph
☐ Chart
☐ Diagram
☐ Complete sentences
☐ Phrase
☐ Filled circle

☐ **E**xamine the information given.
☐ Reread the questions.
☐ Underline key words or phrases.
☐ Think about what the questions are asking.

☐ **S**ee if your answers match the questions.

Step 3

Use the information from "Water on Mars!" to complete the graphic organizer.

Space missions and scientific equipment reveal information about the planet Mars. In the chart below, fill in the information gained from the sources the scientists used to discover if there is water on Mars.

Source	Information Gained
Mariner 4	
Orbiting Probe	
Mars Express and Mars Reconnaissance	
Phoenix Lander	
Balloons	

Step 4

Answer the following questions for "Water on Mars!" using information from your graphic organizer.

1. What kind of information about Mars can an orbiting spacecraft provide?

_____

_____

_____

_____

2. Scientists are designing airplanes and balloons to learn more about the surface of Mars. How will the information these technologies gather about the planet Mars be different from the information provided by an orbiting spacecraft?

_____

_____

_____

_____

_____

3. How did new technology change what scientists thought about water on Mars?

_____

_____

_____

_____

_____

_____

Step 1

Read the scenario
"Characteristics of Scientists."

Characteristics of Scientists

The students were going to the science museum to view an extreme screen movie about the scientist, Jane Goodall. As they prepared for the trip, the students made a list of abilities that are important for scientists. Their teacher added the quality they were describing.

Reasoning—ties to explain observations and say how or why something happens

Insight—tries to figure things out, playing with ideas, looking beyond what is seen with the senses.

Energy—keeps on going

Skill—has knowledge and knows how to use it

Creativity—designs and builds new tools, tests ideas and thinks of how to change and retest them, figures out what makes events or observations happen.

Intellectual Honesty—don't steal ideas, if something doesn't work say so, don't make up data

Skepticism—don't accept something if it does not make sense

Openness to new ideas—willing to accept the fact that your ideas are wrong, can accept evidence if it points against what you think, accept new ideas with evidence.

     © Englefield & Associates, Inc.

The class wrote notes in their journals when they returned to school summarizing Jane Goodall's work.

Jane Goodall–Extreme Screen Movie

Jane Goodall (JG) was born in England. She became interested in chimpanzees when she got a life-size stuffed chimpanzee toy as a gift from her father.

JG worked with chimpanzees in the Gombe Stream Reserve. She lived with the chimps and documented her observations about them. Her documentation and data lead to new information about the chimps.

The chimps had a social organization including:
- their occasional systematic killing of one another and "war-like" situations in the group that JG observed.
- a social nature. JG documented lasting family relationships, courtship patterns, even the adoption of an orphan chimp.
- their hierarchy and their social development. JG's observations included the chimps individual personalities.
- the ability to make and use tools. Until JG documented this, scientists thought that only humans could make and use tools. So scientists started rethinking this.

JG named the chimps instead of giving them numbers like other scientists would do when they studied animals. Scientists were critical about this part of her work.

JG started her work in the Gomeb Stream Reserve in 1960, and she still works for the conservation of chimpanzees in the wild and for better conditions for chimps in zoos and research institutions. There is a plaque honoring JG at Walt Disney World's Animal Kingdom Theme Park.

It is near a carving of the first chimp she named, David Greybeard. It is in the entrance to the "It's Tough to be a Bug!" attraction.

Step 2

Complete the Checklist "Clues for Success."

The checklist will help you to read and think like a scientist.

Clues for Success

☐ **C**arefully read the information.

☐ **L**ook at any illustrations or diagrams.
    They may provide you with additional information to answer the question.

☐ **U**nderstand the way you are asked to answer the question.
    ☐ Graph
    ☐ Chart
    ☐ Diagram
    ☐ Complete sentences
    ☐ Phrase
    ☐ Filled circle

☐ **E**xamine the information given.
    ☐ Reread the questions.
    ☐ Underline key words or phrases.
    ☐ Think about what the questions are asking.

☐ **S**ee if your answers match the questions.

Step 3

Use the information from
"Characteristics of Scientists"
to complete the graphic organizer.

Use the notes in Elizabeth's notebook to show how the characteristics of scientists are reflected in the work of Jane Goodall.

Characteristics of scientists found in the work of Jane Goodall:

Reasoning

_____

_____

_____

Energy

_____

_____

_____

Openness to new Ideas

_____

_____

_____

Step 4

Answer the following questions for "Characteristics of Scientists" using information from your graphic organizer.

1. Why do you think the science community objected to Jane Goodall naming the chimps?

_____

_____

_____

_____

2. Jane Goodall documented her observations about the chimpanzees. What do you think she included in these notes?

_____

_____

_____

_____